Ecocritical Explorations of the Climate Crisis

Ecocritical Explorations of the Climate Crisis expands postcolonial precarity studies by addressing the current climate crisis and threats to the habitability of the planet from a range of ecocritical and environmental perspectives. The collection uses planetary thought-action praxis that acknowledges the interconnectedness of all forms of life in addressing the socioecological issues facing humanity: accelerating climate change, over-exploitation of natural resources, and the Global North–South divide. With reference to contemporary cultural productions, such praxis seeks to examine the ideas, images, and narratives that either represent or impede potential disasters like the so-called sixth extinction of the planet, that inspire the dismantling of carbon democracies arising in the wake of neoliberalism, and that address rising inequality with precarious conditions in the transition to renewable energy. The different chapters explore literary and visual representations of planetary precarity, identifying crisis-responsive genres and cultural formats, and assessing approaches to environment-re/making that call for repair, recovery, and sustainability. In imagining future habitability, they deploy diverse critical frameworks such as queer utopias, zero-waste lifestyles, alternative ecologies, and adaptations to the uninhabitable. The collection tackles problems of global vulnerability and examines precarity as a condition of resilience and resistance through collective actions and solidarities and innovative constructions of the planet's survival as a shared home. It engages with current postcolonial debates, uses intersectional methodologies, and introduces contemporary literary, visual concepts, and narrative types.

Janet M. Wilson is Emeritus Professor of English and Postcolonial Studies at the University of Northampton, UK. Her research focuses on the diaspora and postcolonial writing of the settler colonies of New Zealand and Australia, literature and globalisation, pandemic fiction, transculturalism and transnationalism, and refugee writing. Her most recent publications are the co-edited volume, *New Zealand medievalism: Reframing the* *medieval* (2024) and "Diaspora Screen Media", special issue of the *Journal of Postcolonial Writing* 60.2 (2024). She is editor-in-chief of the *Journal of Postcolonial Writing*, and co-founder and Chair of the global network, Challenging Precarity.

Barbara Schmidt-Haberkamp is Professor of Anglophone Literatures and Cultures at the University of Bonn, Germany. Her main research interests are postcolonial studies and eighteenth-century British literature and culture. She is a member of the steering committee of the international network Challenging Precarity, and has co-edited the collections of essays *Representing Poverty in the Anglophone Postcolonial World* (Bonn University Press, 2021) and *Representing Poverty and Precarity in a Postcolonial World* (Brill, 2022). She is a member of the DFG-funded research training school Gegenwart/Literatur (Contemporary/Literature) and an elected member of the North Rhine-Westphalian Academy of Science, Humanities, and the Arts.

Om Prakash Dwivedi is Associate Professor of English literature at Bennett University, Uttar Pradesh, India. He is the author of *Representations of Precarity in South Asian Literature in English* (2022), *Tracing the New Indian Diaspora* (2014), and co-author with Lisa Lau of *Re-Orientalism and Indian Writing in English* (2014). He is co-founder and vice-Chair of the global network, Challenging Precarity.

Routledge Studies in World Literatures and the Environment

Series Editors: Scott Slovic and Swarnalatha Rangarajan

Decolonial Animal Ethics in Linda Hogan's Poetry and Prose
Toward Interspecies Thriving
Małgorzata Poks

The Imagined Arctic in Speculative Fiction
Maria Lindgren Leavenworth

Desertscapes in the Global South and Beyond
Anthropocene Naturecultures
Edited by Sushila Shekhawat, Rayson K. Alex and Swarnalatha Rangarajan

Reading Madeleine L'Engle
Ecopsychology in Children's and Adolescent Literature
Heidi A. Lawrence

Storying the Ecocatastrophe
Contemporary Narratives about the Environmental Collapse
Edited by Helena Duffy and Katarina Leppänen

Water Stories in the Anthropocene
Anglophone Climate-Change Fiction
Angelo Monaco

Ecocritical Explorations of the Climate Crisis:
Planetary Precarity and Future Habitability
Edited by Janet M. Wilson, Barbara Schmidt-Haberkamp and Om Prakash Dwivedi

For more information about this series, please visit: https://www.routledge.com/Routledge-Studies-in-World-Literatures-and-the-Environment/book-series/ASHER4038

Ecocritical Explorations of the Climate Crisis

Planetary Precarity and Future Habitability

Edited by
Janet M. Wilson,
Barbara Schmidt-Haberkamp
and Om Prakash Dwivedi

Routledge
Taylor & Francis Group
NEW YORK AND LONDON

First published 2025
by Routledge
605 Third Avenue, New York, NY 10158

and by Routledge
4 Park Square, Milton Park, Abingdon, Oxon, OX14 4RN

Routledge is an imprint of the Taylor & Francis Group, an informa business

ISBN: 9781032468792 (hbk)
ISBN: 9781032468808 (pbk)
ISBN: 9781003383598 (ebk)

DOI: 10.4324/9781003383598

Typeset in Sabon
by codeMantra

Contents

Figures

Notes on Contributors

Stefan Benz is Postdoctoral Researcher in the North American Studies Program at the University of Bonn, Germany. His second book project studies how representations of sonic rhythm negotiate Indigenous identity against and beyond ongoing structures of settler colonialism on Turtle Island. He recently co-edited a special issue on "Capitalist Crisis Poetry: Lyric Encounters with Neoliberalism in the Twenty-First Century" for the journal *Amerikastudien / American Studies*.

Wai Chee Dimock is William Lampson Professor of English and American Studies at Yale University. She has published widely on American literature of every period, and is best known for *Through Other Continents: American Literature Across Deep Time* (2007). Editor of PMLA, her essays have also appeared in *Critical Inquiry*, *The Chronicle of Higher Education*, *The New Yorker*, and *The New York Times*. Her team-edited *American Literature in the World: An Anthology from Anne Bradstreet to Octavia Butler* (2017) was published by Columbia University Press. Her monograph, *Weak Planet: Literature and Assisted Survival* (2020), was published by University of Chicago Press.

Om Prakash Dwivedi is Associate Professor of English literature at Bennett University, Uttar Pradesh, India. He is the author of *Representations of Precarity in South Asian Literature in English* (2022), *Tracing the New Indian Diaspora* (2014), and co-author with Lisa Lau of *Re-Orientalism and Indian Writing in English* (2014). He is co-founder and vice-Chair of the global network, Challenging Precarity.

Sonja Frenzel holds a doctorate in contemporary English literature and has completed a second academic book on artistic research in literary studies. She currently works as a writer and freelance translator from English and French. Her work attempts to capture the embodied experience of presence and precariousness. It is inspired by ecology and eco-translation, gender and queer studies, as well as by modern aesthetics and philosophies of modernity. Her most recent projects include the multi-lingual translation of Vincent Broqua's poetry collection *Recovery* and her own novella "born later … (a response to B. Brecht)".

Cristina M. Gámez-Fernández is Senior Lecturer of English in the Department of English and German at the University of Córdoba (Spain) and a founding member of the Challenging Precarity network. She has recently co-edited *Representing Vulnerabilities in Contemporary Literature* (2023), *Embodied Vulnerabilities in Literature and Film* (2023), and *The Spanish and Latin American Legacy in North American Poetry and Art* (2024).

May Joseph is Professor of Social Science and Cultural Studies at the Pratt Institute, where she teaches a walking history of coastal New York. Joseph is founder of Harmattan Theater and has produced site-specific performances along Dutch and Portuguese maritime routes. Joseph's publications include *Aquatopia: Climate Interventions* (2023), co-authored with Sofia Varino; *Sealog: Indian Ocean to New York* (2019); *Ghosts of Lumumba* (2019); *Fluid New York: Cosmopolitan Urbanism and the Green Imagination* (2013); and *Nomadic Identities: The Performance of Citizenship* (1999). mayjoseph.com

Chiara Lanza is a PhD candidate in Sustainable Development and Climate Change at the University Institute of Higher Studies of Pavia (IUSS) and University of Ferrara, Italy. By bringing the soil sciences and literary studies together, her research project aims at analysing representations of more-than-human entanglements, environmental injustice, and soil degradation in works of literature by contemporary Anglophone and Francophone authors. While rooted in material ecocriticism, animal and plant ethics, and decolonial studies, this approach reaches out to the sciences to highlight the ways in which texts speak across disciplines, geographies, and (hi)stories, thus making environmental dynamics emerge. Her research interests range from acts of gardening in the Caribbean to the legacies of nuclear colonisation in the Pacific Ocean.

Klara Machata is a PhD candidate and Research Associate in English Literary and Cultural Studies at the University of Freiburg, Germany. Her doctoral dissertation is concerned with space, place, and the Anthropocene in contemporary Anglophone literature with a focus on South Asia and Southeast Asia. Her research and teaching are informed by her studies in geography and English literature and explore the intersection of several disciplines within the environmental humanities: ecocriticism, postcolonial theory, and political and cultural geography.

Pramod K. Nayar teaches in the Department of English, the University of Hyderabad, India, where he also holds the UNESCO Chair in Vulnerability Studies. Among his newest books are *Vulnerable Earth: The*

Literature of Climate Crisis (Cambridge, 2024), *Nuclear Cultures: Irradiated Subjects, Aesthetics and Planetary Precarity* (Routledge, 2023), *The Raj: A Journey through Ten Documents* (Bloomsbury, 2023), *Alzheimer's Disease Memoirs: Poetics of the Forgetting Self* (Springer, 2022), *The Human Rights Graphic Novel: Drawing It Just Right* (Routledge, 2021), among others. He is an elected Fellow of the English Association and the Royal Historical Society.

Jan Rupp is currently Coordinator of the Research Centre for the Study of Culture (RCSC), Giessen, having served as an interim professor at the universities of Frankfurt, Giessen, Heidelberg, and Wuppertal. He is the author of *Genre and Cultural Memory in Black British Literature* (2010) and a second monograph on representations of ritual in modernist *Pageant Fictions* (2016). His research interests include the contemporary novel, cultural memory studies, narrative theory, intermediality, ritual in literature, and (neo-)Victorian studies. He has published widely on diasporic British as well as postcolonial Anglophone writing and theory, including on didactic perspectives of the literature classroom. Among his current work is a project on figurations of world writing and environmental memory in literatures of the Global South.

Barbara Schmidt-Haberkamp is Professor of Anglophone Literatures and Cultures at the University of Bonn, Germany. Her main research interests are postcolonial studies and eighteenth-century British literature and culture. She is a member of the steering committee of the international network Challenging Precarity and has co-edited the collections of essays *Representing Poverty in the Anglophone Postcolonial World* (Bonn University Press, 2021) and *Representing Poverty and Precarity in a Postcolonial World* (Brill, 2022). She is a member of the DFG-funded research training school Gegenwart/Literatur (Contemporary/Literature) and an elected member of the North Rhine-Westphalian Academy of Science, Humanities, and the Arts.

Leonor María Martínez Serrano is Associate Professor in the Department of English and German Philology, Faculty of Art and Humanities at the University of Córdoba, Spain. Her research interests include Canadian and American Literature, High Modernism, and Ecocriticism. She has co-edited *Modern Ecopoetry: Reading the Palimpsest of the More-than-Human World* (Brill, 2021) and authored *Breathing Earth: The Polyphonic Lyric of Robert Bringhurst* (Peter Lang, 2021).

Scott Slovic retired as the University Distinguished Professor of Environmental Humanities at the University of Idaho, USA, where he was teaching since 2012—previously he was a professor at Texas State University and the University of Nevada, Reno. He served as founding president of the Association for the Study of Literature and Environment (ASLE) from 1992 to 1995, and since 1995 he has edited *ISLE: Interdisciplinary Studies in Literature and Environment* for ASLE and Oxford University Press. He is the author, editor, or co-editor of 27 books, including, most recently, *The Bloomsbury Handbook to the Medical-Environmental Humanities* (co-edited with Swarnalatha Rangarajan and Vidya Sarveswaran). His forthcoming books include *Nature in Literary Studies* (co-edited with Peter Remien) for Cambridge University Press's Critical Concepts Series.

Sofia Varino is an author and lecturer whose work focuses on transdisciplinary ecologies. Varino is co-author of *Aquatopia: Climate Interventions* (2023) and co-editor of a special issue of *Somatechnics* on "Data Matters" (2019). Varino has published in leading academic journals including *Catalyst*, *SHIMA*, *Somatechnics*, *European Journal of Women's Studies*, and *Women's Studies Quarterly*. Varino holds a PhD in Comparative Literary and Cultural Studies from SUNY Stony Brook. sofiavarino.com

Aleks Wansbrough, based in NSW, Australia, lectures at the Sydney College of the Arts, University of Sydney, and at Bennett University, India, where he is Adjunct Professor. The author of *Capitalism and the Enchanted Screen: Myths and Allegories in the Digital Age* (2021), his current research concerns ideology and contemporary culture. He is an editor of the *Journal of Asia-Pacific Pop Culture*, and co-editor of the journal, *Alterity Studies and World Literature*.

Janet M. Wilson is Emeritus Professor of English and Postcolonial Studies at the University of Northampton, UK. Her research focuses on the diaspora and postcolonial writing of the settler colonies of New Zealand and Australia, and on literature and globalisation, transculturalism and transnationalism, and refugee writing. Her most recent publications are the co-edited volume, *New Zealand Medievalism: Reframing the Medieval* (2024) and "Diaspora Screen Media", special issue of the *Journal of Postcolonial Writing* 60.2 (2024). She is editor-in-chief of the *Journal of Postcolonial Writing*, co-editor of *Studies in World Literature* (Ibidem-Verlag), and Chair of the global network, Challenging Precarity.

Kanak Yadav teaches English in the School of Liberal Arts, IIT Jodhpur, India. She is also affiliated with the Interdisciplinary Research Platform (IDRP), Digital Humanities, at Indian Institute of Technology, Jodhpur. She completed her PhD at Jawaharlal Nehru University, New Delhi. She has published in the *Journal of Postcolonial Writing*, the *Journal of Commonwealth Literature*, the *Contemporary Voice of Dalit, Akademos*, and *World Literature Today*. Her research interests include literature and cities, Dalit narratives, Indian English novels, postcolonial anglophone fiction and environmental humanities.

Acknowledgements

The editors would like to thank the Routledge series editors, Michelle Salyga and Bryony Reece for their understanding and support throughout this process, and Sophia Ahmad, Marie Berndt, and Lena Falk for their help in preparing the manuscripts.

Janet M. Wilson, Barbara Schmidt-Haberkamp, and Om Prakash Dwivedi

Introduction

The Anthropocene and the "weak planet": framing postcolonial precarity ecocriticism

Janet M. Wilson, Barbara Schmidt-Haberkamp and Om Prakash Dwivedi

Environmental precarity is at a critical juncture in the current era of the Anthropocene, infiltrating the popular imaginary and becoming accepted as a feature of everyday life. Extreme weather events, the melting of the polar ice caps, resource depletion, loss of biodiversity, and species extinction point to an unpredictable, unstable planetary future. Jeopardizing of the planet's health and hence the continuity of human and nonhuman life has become the systemic crisis of the 21st century, of which the recent Covid-19 pandemic might be seen as a foretaste. Covid-19 magnified awareness of catastrophe and the "sense of universal vulnerability" (Slovic 2020, n.p.). As a global epidemiological emergency that heightened inequality, insecurity and risk, it can be linked to ecological attrition as imagined in futuristic dystopias of science fiction such as further loss of habitat and species extinction, so providing the zoonotic conditions for interspecies exchange and the proliferation of new pathogens. At the same time, weather catastrophes create populations of climate refugees fleeing from destruction of their territories and homelands. The prospect of such widescale disaster raises concerns about what Rob Nixon (2014, n.p.), referring to "unequal human agency, unequal human impacts, and unequal human vulnerabilities", calls "the Great Divergence". While the superrich may be protected, the impacts for vulnerable others at the bottom of the socioeconomic scale will increase exponentially, as Upendra Baxi (2020, n.p.), writing on Covid-19 in India, comments: "If the spread of a global pandemic was an act of misfortune, its catastrophization has to be located in the injustices of social structure and policy". In the Global South in particular, social and economic precarity caused by "relatively unstable and dispersed conditions of deprivation and insecurity" (During 2015, 58), is likely to be exacerbated.

As well as uneven impacts, global warming has an irreversible momentum, from the extremes of sudden devastation caused by catastrophic weather events – cyclones, bush fires, droughts and floods – to the slow

DOI: 10.4324/9781003383598-1

attritional violence "dispersed across time and space [that] typically is not viewed as violence at all" (Nixon 2011, 2). The massive environmental upheaval caused by the extractive practices of the fossil fuel industries – acidification of oceans, chemicalization of life, deforestation, melting of the cryosphere – has reached unprecedented levels today. Earth system analysis recognizes that these regional and global instabilities are developing beyond safe operation spaces, as key climate tipping points have been reached, such as the melting of the ice sheets in the Arctic and Antarctic, and destabilization of the Amazon rain forest (Armstrong McKay et al. 2022, n.p.; Lenton et al. 2023). Most recently (April 9, 2024), a leading scientist, Dr. Gavin Schmidt, director of NASA's Goddard Institute for Space Studies, has said we are entering "uncharted territory" as the last ten months (May 2023–March 2024) have seen unprecedented new levels of warming caused mainly by the weather pattern of El Niño and climate change (McGrath and Poynting 2024, n.p.).

Scientific knowledge about these impacts drawn from continued observation of climate processes and interactions has informed globalized systems of governance in their attempts to sustain resource systems and stabilize the climate (Oldfield et al. 2014, 4). Constant monitoring of environmental change has helped shape the policy direction of nations towards mitigation of global warming in recent years, with landmark decisions reached in the annual UN Climate Change conferences, COP29 in Baku in 2024 being the most recent, even though there is dissent over the formal agreements and a shortfall in implementing recommendations (Dwivedi 2023, 4). Most notably, in the transition towards carbon neutral societies, fast spreading contagious processes of social and technological change are being introduced to disrupt the planetary socioeconomic system. Called social tipping interventions (STIs), these aim for "social tipping elements" (STEs), and reductions in anthropogenic greenhouse gas emissions, for example, by removing fossil fuel subsidies, building carbon neutral cities and mobilizing decentralized energy generation (Otto et al. 2020, 2354). Other global interventions appear in the revised conservation agenda of biodiversity science aimed at restoring ecological functions through processes of rewilding (Jepson and Blythe 2020, 8, 156). Important for this volume, which aims to make an intervention in a range of interdisciplinary fields and to develop the concept of postcolonial precarity ecocriticism, are the expanded concepts of precarity, vulnerability, resilience, and resistance in relation to socio-ecological systems developed in the arts, humanities, and social sciences. The new disciplinary syntheses encourage wide-ranging approaches to planetary survival and habitability (Oldfield et al. 2014, 6).

These responses to the planetary emergency acknowledge that more extensive collaborations are needed. *Ecocritical Explorations of the Climate Crisis* takes its ethical and critical bearings from the multiple intersections between the fields of postcolonialism and ecocriticism that have shaped an environmental postcolonialism, and triangulates them with theories and debates about precarity that are informing research in the social and political sciences – for example, in social geography and migration studies – as well as in literature and cultural studies (Schmidt-Haberkamp and Gymnich 2022, 6; Ashcroft, Griffiths and Tiffin 2006, 6). As Graham Huggan and Helen Tiffin ([2010] 2015, viii) argue, "global warming [...] requires a broader *ecological* conception of natural-cultural relations on several different levels – including the planetary level – than has tended to be the case in much ecocriticism until now" (original emphasis). Achille Mbembe (2019) demands decolonization on a planetary scale to "reconstruct the world in common" and to "reinvent forms of life in common that go beyond the requisite of the nation state, ethnicity, race, religion" (n.p.). Theorization of a "radical decolonial" approach informs Stuart Cooke's and Peter Denney's (2021) critical framework of transcultural ecocriticism (2), reinforcing their claim that the "critical project [of reading, interpreting and translating data and texts on the planetary emergency] has become planetary, biological, even geological" (1). In the same vein, Pramod K. Nayar (2024), in *Vulnerable Earth: The Literature of Climate Crisis*, issues a clarion call "to rethink species boundaries, and examine the possibilities of multispecies becoming and belonging" (19).

More complex and interconnected conceptual frameworks of analysis have generated transdisciplinary expansion between the natural sciences, social sciences and humanities (Oldfield et al. 2014). In the environmental and energy humanities, the tradition of ecocriticism has expanded its boundaries and developed sub-disciplinary intersections with areas such as postcolonial ecology and ecocriticism, indigenous ecocriticism, transcultural and empirical ecocriticism, and, as this volume showcases, postcolonial precarity ecocriticism. Such interdisciplinary interconnections and dialogues have helped scale up the research with the synthesizing that the scale of the crisis demands. The rise of fields such as the blue humanities (Hofmeyr 2019), the energy humanities (Szeman and Boyer 2017; Mišík and Kujundžić 2021) and the "ethical practice of 'becoming-witness'", that is, of cultivating attentiveness to "the diverse living beings and forms of liveliness that constitute our world" (Rose and Van Dooren 2017, 120), are playing a crucial role in mapping the historical stages and processes that have led to the present crisis, and in the broadening of conceptual frameworks.

Across all disciplines, study of the environment has been transformed by a sense of moral urgency since the formal recognition of the Anthropocene as a geological era in 2000 by Nobel laureate atmospheric chemist Paul J. Crutzen and ecologist Eugene F. Stoermer (2000). The Anthropocene refers to the damaging impact of human activity on the planet and erosion of the

Earth's system due to mankind's "central role in geology and ecology" (17), but more commonly the term is understood as demarcating a new epoch, one that has advanced spectacularly since the post-1945 "Great Acceleration" of consumerism, industrialization and technology (Nixon 2014, n.p.). The acknowledgement of human geological agency has undermined the very idea of the human and other accepted categories, breaking down the human-nature dichotomy and demanding reconceptualization of human-environment interactions. The very concept "anthropogenic climate change" signals a necessary reconfiguration of human agency: it has generated a need for greater accountability and responsibility as efforts intensify to find solutions to the planetary emergency. The awareness that humans can shape the geological future has driven new ethical concerns and ushered in a renewed political commitment to mitigate negative planetary impacts.

While the Anthropocene is understood as defining the current epoch, the related concept of the Capitalocene, introduced by sociologist and environmental historian Jason Moore (2015, 2016), looks to the historical origins of the climate crisis in empire and capital. Reading the current planetary emergency through economic and historical rather than geological time frames, Moore argues the "Capitalocene signifies capitalism as a way of organizing nature – as a multispecies, situated, capitalist world economy" (2016, 6). He refers to the predatory system of capital and the fossil fuel economy that creates an entanglement of society and nature, tracing this to as early as the 15th century, when humans began to control and exploit nature through a process of massive deforestation (2015, 145). As one source for the environmental logic of late capitalism, the Capitalocene provides a political, economic narrative that casts the blame for climate change on the system of capitalism, according to Timothy Clark (2015, 101): i.e., the captains of industry, governments and economic planners.

The recognition that not all humans share equal responsibility for the crisis cuts across the universalist assumption of an undifferentiated humanity implicit in the notion of the Anthropocene, even though mankind's collective agency and shared responsibility appear in concepts such as the carbon footprint (Chakrabarty 2021, 57–58). But as Dipesh Chakrabarty (2021) points out, in linking the environmental crisis to the history of capital and interpreting it as a crisis of capital,

> [W]e lose sight of the nature of our present that is defined by the coming together of the relatively short-term processes of human history and other much longer-term processes that belong to Earth systems history and the history of life on the planet.
>
> (58)

These processes that preceded mankind's invasive presence and may still be at work in the current crisis should not be overlooked. Yet, the current acceleration of the anthropogenic environmental crisis demands committed responses on a global scale. Postcolonial theories that challenge exploitation and oppression, can extend to the massive social and economic inequities associated with the uneven impacts of global warming, and they target as root cause an economic system that combines the history of global capitalism with present-day neoliberal market forces. Postcolonial ecocritic Johan Höglund (2020), for instance, addresses the precariousness of populations and environmental ecologies, when he asks how far global capitalism is prepared to adapt in the face of such inequality and to provide greater security for large sections of the world's population struggling under the new pressures to sustain the world economy and recover planetary habitability (450).

The essays in *Ecocritical Explorations of the Climate Crisis: Planetary Precarity and Future Habitability* respond to the crisis of a vulnerable, fragile "weak planet" (Dimock 2020), or a "planet in peril" (Nayar, this volume), and offer broadly conceived considerations of alternative perspectives, solutions or measures of repair and agency that might increase the Earth's habitability as a shared home. Situated in the disciplines of the humanities and social sciences, the volume engages with debates on pandemics and epidemiology, climate injustice, consumer culture, decolonization, art, literature and performance as forms of activism, and the posthuman. The 13 chapters use diverse theoretical and critical approaches such as queer utopias, zero-waste lifestyles, hydrosophy, planetary dysphoria, and alternative indigenous ecologies, and they draw upon an eclectic array of text types and genres – the graphic novel, speculative fiction, poetry, feature films, installations, TV series and documentaries, blogs and websites, oral songs and performance praxis – in order to imagine future conditions and systems needed for planetary sustainability.

Underlying this text-based methodology are assumptions about the importance of imagining planetary habitation, such as those voiced by Richard Kerridge and Neil Sammells (1998), namely that texts can be platforms for articulating the complicated nature of the relationship between human beings and their immediate environment. Indeed, several contributors to this volume – Klara Machata, Jan Rupp, Scott Slovic, Chiara Lanza, and Aleks Wansbrough – pivot their arguments explicitly around the claim that literature and art's affective, imaginative and self-reflective powers can give voice to different aspects of climate change, such as exposing the cultures of concealment associated with big business or urging critical reflection upon disaster, in ways that earth science and systems analysis cannot.

In this way, the volume challenges the assertion of critics like Timothy Clark (2015, 187) that literature is unable to deal with the multi-scalar challenges of the planetary, or Emily Apter's (2013) concept of "planetary dysphoria" and the stalling of narrative due to ecological collapse, or Amitav Ghosh's (2016) view that global warming is resistant to the arts (73), because adequate narrative forms do not yet exist. The different contributions draw on a range of socioeconomic phenomena associated with climate change within which their texts are grounded – the "material conditions that both enable and help textual production" (Carrigan 2011, 22). These both reflect and anticipate wider social, political, and cultural shifts: combined with imaginings of more hopeful futures they construct a social imaginary of planetarity as a relational, multi-scalar world system, and they offer images of the survival of the planet as a shared home.

Diverse forms of textual and visual representation are invoked to address planetary precarity and offer possible resolutions or alternative story worlds in narrating the impacts of environmental degradation and related histories of exploitation and oppression. Among these interventions are the forms of healing, resilience and resistance that can envisage new futures and utopian worlds. Literary and visual texts interrogate and unsettle the binaries that have shaped western imaginations, geographies, and institutions, such as human-nonhuman, nature-culture, urban-rural, as they imagine entanglements between "different life forms, time frames, spirits and realities" (Wilke 2019, 144; see also Cooke and Denney 2021, 3, 13). They identify and interrogate the affective dimensions of precarity and vulnerability found, for instance, in the reinforcement of ties to everyday routines and commonplace acts. Drawing upon indigenous epistemologies, traditional ecological knowledge, and grassroots movements, they consider alternatives to neoliberal practices of placemaking and dwelling aimed at planetary recovery, and "endorse the solidarities and collectivities associated with environmental and refugee activism" (Wilson, Dwivedi, and Gámez-Fernández 2020, 442).

In this range of responses emerging from the intensified interdisciplinarity of climate change research, the collection presents a hybrid disciplinary framework of postcolonial precarity ecocriticism. Aiming to interrogate the multiple and overlapping global, decolonial and ecological processes underlying this relational field, the chapters come together through a theoretical reliance on the concepts of precarity developed by Judith Butler (2004; 2009), integrating their framework with the practices and arguments of postcolonial studies and ecocriticism. This disciplinary synthesis reflects the polysemy of the Anthropocene as a category situated between geology and earth sciences (Parham 2021, 9; Chakrabarty 2021 3), and, in exploring new modes of habitability, it facilitates scholarly engagement with the vastly different scales of history, space and time that constitute a major challenge of the planetary.

Postcolonial precarity ecocriticism intersects environmental issues with precarity, race and colonialism. Its decolonial politics identifies historical colonial dispossession in the current entangling of neoliberal capitalist practices with environmental degradation, and it creates dialogue with theories of precarity and vulnerability concerned with inequality and injustice on the one hand, and ecocritical discourses that emphasize closer human/nonhuman affiliations, and urge respect for all living entities on the other. As an ecological enquiry this critical field argues that the Anthropocene narrative of man's geological force underlying the current climate emergency is a legacy of colonization and the exploitation of territory and labour during the expansion of empire (Poray-Wybranowska 2021, 13). In keeping with the historical contexts of the Capitalocene, postcolonial precarity ecocriticism identifies (neo)colonial injustices inherent in the extractive and rapacious exploitation of the environment with the powers responsible for Black slavery and other forms of oppression that render human and nonhuman lives precarious. It recognises with critic Julia Martin (2017) that "the othering and exploitation of ecosystems and nonhuman beings is inseparable from the othering of people, that the violence of resource extraction is linked with violence against women, that the war against the Earth is a war against ourselves" (177); and it uses decolonial, ecocritical frameworks to represent the parallels between race and gender, inequality and abuse in the colonial past, and in neocapitalist agendas. As Fiona Moolla (2016), in writing on African ecocriticism, points out "the environment cannot be defended without engaging colonialism, capitalism and imperialism" (8), while decolonizing the imperial ecological aesthetics associated with "Western modes of conquest, categorization, and extraction" is the aim of transcultural ecocriticism (Cooke and Denney 2021, 2). Its focus on First Nation ecologies and indigenous perspectives can also be found in the chapters in this volume by Wai Chee Dimock, Jan Rupp, Chiara Lanza and Leonor Serrano.

Overlapping with postcolonial ecocriticism's challenge to unequal power relations and social injustice, precarity in the sociological sense refers to the uneven development of economic world processes due to neoliberal hegemony over production. The politics of precarity recognize that as a systemic process neoliberalism's roots lie in colonialism with its historic patterns of exploitation. For "if colonialism was symptomatic of increasingly uneven power structures, then neoliberalism can be seen as replacing it, albeit in new forms and shapes" (Wilson, Dwivedi, and Gámez-Fernández 2020, 441). Greater flexibilization of employment practices and lack of protection for workers define the "post-welfare, post Fordist neoliberal state" (Lesutis 2022, 23). As Simon During says, "as the politics of subalternity were largely absorbed into the machinery of emergent neo-liberal state capitalism", labourers and workers were converted into the precariats, or denizens of the precarious society (2015, 57). Yet, these unstable labour conditions belong to a wider structure of exploitation and dominance, one that has expanded geographically beyond the traditionally dominant Global South/Global North divide, as evidenced by the pockets of precarious lives in the midst of affluent western societies (Schmidt-Haberkamp and Gymnich 2022, 3).

In their philosophical enquiry Judith Butler distinguishes between precariousness that is an ontological condition of human life in that all subjects are bound by their vulnerability, although some are more exposed to illness, war, disease, and famine than others, and precarity as the type of dependency that they call a "politically induced condition in which certain populations suffer from failing social and economic networks of support more than others, and become differentially exposed to injury, violence, and death" (2004, 25). The recognition of precarity in conditions that cut "across identity categories as well as multicultural maps" (32) has become increasingly relevant in understanding the widening inequalities in the economic world system, and has expanded into a more complex discourse in which cognate terms like uncertainty, insecurity, vulnerability and fragility point to a more active engagement through practices aimed at recovery and repair. Recent manifestations of precarity such as the current upsurge of refugees and asylum seekers desiring better opportunities or escaping persecution, like the urban self-precaritization of young people, and the financial insecurity of the gig economy involving heightened risk for casual and part-time workers, are components of the "new, dangerous" society (Standing 2011). Such material evidence of hardship, however, goes some way to support postcolonial critiques of Judith Butler's abstract humanist ethics, like those of Sunera Thobani (2010) and Ida Danewid (2017). Danewid asks whether "an ethics grounded in a generalized suffering" is "the best way of encountering contemporary forms of violence and injustice" (1677), claiming that it erases imperialist history, the fact that the "shared global present [is] built on colonialism, racism, and white supremacy" (1683), and so "transforms the responsible colonial agent into an innocent bystander" (1684). Thobani argues that Butler's arguments reflect a "western supremacy" that is rooted in colonialism and racism and marginalizes precarity's systemic nature (2010, 135).

Precarity overlaps with ecocriticism through its reference to modes of resistance and resilience that might bring about planetary repair, and a concerted response to climate injustice, as found in Butler's concept of "precarity as activism"; that is, of challenging those who exploit or wish to dispose of populations (2015a, ix). It has affinities with ecocriticism's advocacy of respect for all living and sentient beings, for affirming close bonds between human and nonhuman by way of protecting and caring for the environment. Pramod K. Nayar (2019) acknowledges that the climate crisis and precarity are inextricably interconnected in introducing the term "ecoprecarity", which implies that planetary change cannot be viewed in isolation and that the multispecies relatedness with nature is essentially symbiotic (9). As a significant framework for thinking through ways of being in the world, ecoprecarity addresses the entanglement between humans, nonhuman living beings, and the changing environment, for it "is at once about the precarious lives *humans* lead in the event of ecological disaster [...] and also about the environment itself which is rendered precarious due to human intervention in the Anthropocene" (7; original emphasis).

As these semantic mutations and nuances of meaning arising from the environment crisis indicate, precarity has become a highly applicable concept in social, economic and political fields. As a term of literary and cultural enquiry qualified by categories of analysis such as race, class and gender, it dynamically intersects with discourses on the Anthropocene in both the arts and sciences, as Wai Chee Dimock shows in her chapter on oxygen and our "precarious breath" as grounds for creativity. The plasticity of the category of precarity in adapting to diverse disciplines and its ubiquity in debates about changing conditions in social, economic and planetary spheres mark out its importance for the present moment, even to the point of overuse as Schmidt-Haberkamp and Gymnich point out (2022, 1, 7). Judith Butler claims, it is "a new form of regulation that distinguishes this historical time" (2015a, vii); while Anna Tsing (2015) asks: "What if precarity [...] is the condition of our time – or, to put it another way, what if our time is ripe for sensing precarity?" (20). A byword for diminished scale, for hollowing out and residues, it stakes a claim for the dark side of the history of globalization, including colonial oppression and exploitation, and the human and more-than-human costs of neoliberalism as found in the fragile, unpredictable climate, and the recent pandemic whose disruptions echo those occurring in the biosphere. Yet precarity also reveals the dualism of human agency, as Wai Chee Dimock points out in her chapter, as both harm-bearing as well as harm-mitigating in its capacity for healing and well-being: in championing relationality and interspecies connectedness, then, precarity is an energizing power in mobilizing the search for alternatives to neoliberal sources of knowledge and solutions, in terms of recovery, resilience and repair.

Ecocriticsm, the planetary and the world of the text

The Anthropocene is a threshold concept in terms of scale according to Timothy Clark writing in 2015, and Chakrabarty concurs, saying, "we live on the cusp of the global and the planetary" (2021, 204), differentiating between the "global" as a "humanocentric construction" and the "planet" that "decentres the human" (19), and he demands that we consider its habitability for the various life-forms existing on it (87). According to Chakrabarty, we are currently

> passing through a unique phase of human history when, for the first time ever, we consciously connect events that happen on vast ecological scales – such as changes to the whole climate system of the planet – with what we might do in the everyday lives of individuals, collectivities, institutions, and nations.
>
> (2021, 155)

The uncertainty caused by planetary disorder on such a vast scale and manifested in unpredictable climate events is, according to Marco Caracciolo (2021, 8), on a higher level than other types of existential doubt or insecurity, and so poses a greater challenge to the imaginative and fictional capacities of the critic and artist, as well as to readerships and audiences. Astrid Bracke (2018) points out that new fictional constructions and analytical frameworks are needed, and a different kind of Anthropocene fiction is called for to deal with anthropogenic climate change. In his 2015 study, *Anthropocene Fictions*, Adam Trexler has created a framework in order "to gather, historicize, and interpret climate change novels" (16; see also Parham 2021). Aleks Wansbrough in his chapter asserts the importance of narrative for encountering the climate emergency as a medium in which to wage the struggle for habitability because "human and political agency is moored in narrative". Trexler goes so far as to suggest that novels can have agency (Wallace 2017, 567–568, citing Trexler 2015, 16), although the question of whether an activist literary ecocriticism is possible, as asked recently by John Parham (2023, 1) remains a challenge (Wallace 2017, 568).

Other contributions to this volume also consider ways of going "beyond articulation" (Wallace 2017, 574) to engage with the issues of effective representation, agency and activism in asking how different text forms can work effectively to halt climate change or bring about planetary repair. Jan Rupp, for example, argues that poetry is better equipped than the novel to express the complexity and multiplicity of environmental issues and to offer new solutions; Sonja Frenzel advocates a form of eco-translation, and a "literary tradosphere" (a literary planeary translation system) in order to think both backward and forward towards more-than-human storytelling; while May Joseph's and Sofia Varino's performance praxis on ocean sites and landscapes combines reference to the deep, geological history of oceans with affective memories of water against landscape. Dimock identifies Native American resilience during the Covid-19 pandemic in the creative practice of Muscogee (Creek) poet and musician, Joy Harjo, and finds reparative agency arising from the injuries of the past.

These approaches suggest that the essays in this collection can be productively contextualized in relation to the arguments of Timothy Clark in *Ecocriticism on the Edge: The Anthropocene as a Threshold Concept* (2015) and Amitav Ghosh in *The Great Derangement* (2016), namely that the "derangement of scales" (or scale effects, i.e. the impacts of translating between the scales of the planet and the scales of the local) prevents writers and novelists from actively engaging the problem and proposing solutions; instead, they move towards ideas of mitigation or even avoidance. Thus, the Anthropocene is also a crisis of representation, as Rob Nixon has argued:

> how can we convert into image and narrative the disasters that are slow moving and long in the making, disasters that are anonymous and that star nobody, disasters that are attritional and of indifferent interest to the sensation-driven technologies of our image-world? How can we turn the long emergencies of slow violence into stories dramatic enough to rouse public sentiment and warrant political intervention?
>
> (2011, 3)

Ghosh has famously contended that the realist novel, accustomed to representation of social phenomena, cannot cope with the unexpected and the challenge posed by the vast scales of the planetary (2016). Clark argues that the fiction writer prioritizes the human scale and that multidimensional complexities on the planetary or interstellar scale cannot be accessed beyond the ordinary faculties of sight and comprehension. "Scaling up" in Clark's terms requires registering the ecological footprint (2015, 9–11), that is, multiplying to a higher degree individual incidents of climate abuse (as with air travel) demands a reconfiguration of complicity and responsibility on a greater scale than the individual one. At the same time, scaling up of the human imagination serves to emphasize the anthropocentrism of the Anthropocene, as Cooke and Denney remind us (2021, 8), even though the inclusion of nonhuman perspectives and advocacy of a more heterogeneous response (rather than a focus on a single genre or region) are vital in representing the magnitude and diversity of planetary precarity.

Multiplicity and diversity in temporal and spatial terms are broadly seen as essential for conceptualizing a new aesthetic that takes into account the scale of the planetary. Clark advocates a scale critique that aims to find larger assemblages through defining patterns of relation, and according to John Parham this lays the ground for the "adaptive, hybrid form of literature that the Anthropocene will necessitate" (2021, 7). An example comes from Klara Machata who, in her study of the novella "Entanglement", finds an intersecting network of humans and nonhumans in multiple interwoven stories that "seem to be written from deep within the socio-ecological entanglements of a planetary age"; another comes from Jan Rupp's reading of the performance ecopoetry of Marshall Islands poet, Cathy Jetnil-Kijiner, to show that poetry's multiperspectival dimensions enable a grasp of the scalar complexity associated with the planetary.

Speculative fiction is one of those genres dismissed by Ghosh, as it exists on the peripheries of realism with its attendant paradigms of social responsibility and conscience, but in fact its futuristic and apocalyptic orientation has always been congenial to representing radical fantasies of planetary or interstellar activity and disorder, so justifying Gil Hochberg's (2018) claim that its "dark futurism and dystopic imagination" may signal ethical concerns, and generate eco-awareness of the planet's future (35). This can be aligned with Clark's notion of scale framing (by which to establish spatial and temporal boundaries to consider phenomena and identify patterns), as well as scale effects – for the discontinuities of larger scales in which things happen differently create "cognitive estrangement" or dissonance (Parham 2021, 17). Affective responses to speculative science fiction about topics such as the future colonization of different worlds, to apocalyptic climate fiction set on inhospitable planets, or indeed on an uninhabitable planet Earth, as discussed by Cristina M. Gámez-Fernández in her chapter, and cognate genres such as petrofiction about the extractivist practices of neoliberal capital, which Chiara Lanza addresses, create denial, ambivalence, horror – impacts that evoke the alienation and shock associated with the disjunctiveness of scales.

Overview

In the first section on "Planetary precarity and vulnerability", the challenges posed by human and nonhuman precarity are considered alongside collaborations and initiatives aimed at new forms of agency, survival and habitability. The section's ethical orientation can be read in terms of Judith Butler's (2004) questions in the wake of the Twin Tower attacks of 9/11 about the precariousness and grievability of some lives rather than others, and their advocacy of making precarious, diminished lives more visible. The three chapters give representational space to vulnerable human and nonhuman beings and locations affected by climate change or nuclearization, and their responses and proposed solutions.

In their focus on planetary damage and repair they draw on the concept of vulnerability, a term that, as Cristina M. Gámez-Fernández and Miriam Fernández-Santiago (2023) explain, has recently expanded semantically to encompass a social ethics that recognizes the need for care (2–3). During Covid-19, this heightened recognition catalysed calls for state intervention and provision in acknowledgement of the collective responsibility and ethics of care required in a pandemic (Fineman 2021, 184,

190; Michaels 2021, 755). Like precarity, the experience of vulnerability between individuals, collectivities, communities and nations (Butler 2004, xii; 2009, 25) is inconsistent, with some being more vulnerable and affected than others. In this section, vulnerability's uneven spread geographically and socially, and its dualism as a concept, as both wounding and caring, passivizing and agentive (Gámez-Fernández and Fernández-Santiago 2023, 3–4), are emphasized through contrasting textual representations: either positively as an enabling condition of a resistant or resurgent resilience, as Wai Chee Dimock illustrates, or negatively, drawing on its traditional connotations of marginality and victimhood, as involving exploitation and manipulation, as Cristina M. Gámez-Fernández shows.

The inherent contradictions in the concept of vulnerability underpin Judith Butler's theorizing of the body's social vulnerability as implying "morality, vulnerability, agency" (2004, 26). Butler considers vulnerability as involving a form of affective relationality based on "the subject's radical dependency and capacity to affect and be affected", as Leticia Sabsay (2016, 279) points out, and this is "enmeshed with the capacity to act" (285). That is, vulnerability can be delinked from connotations of victimhood and generate forms of agentic resistance, because by being neither fully active nor passive (Butler 2016, 25–26) it occupies an indeterminate position between being affected and acting. Butler's point that human-nonhuman vulnerability is "an effective mobilising force" (2016, 14) in demonstrations and assemblies that enact public will, and her *Performative Theory of Assembly* (2015b) provide a framework for assertions and enactments of political resistance in this volume – as in Stefan Benz's analysis of a socially differentiated social precarity in *Escaping Exodus*, in May Joseph's and Sofia Varino's discussion of the experimental environmental performances of Harmattan Theater as bodily assemblies, and in Pramod K. Nayar's overview of the 20th-century anti-nuclear protest movement.

Each chapter in this section opens with an image of planetary frailty due to global catastrophe: the Covid-19 pandemic, planetary nuclear precarity, and a fictional postapocalyptic wasteland. The scale of disaster can be inferred from the eclectic range of text forms: epidemiological documents, indigenous oral poetry and sound, the visual and written media of anti-nuclear protest, and diverse genres of postapocalyptic climate fiction. In the opening chapter, Wai Chee Dimock reflects on oxygen, the elixir of life, and the precarity of breath as in the call "*I can't breathe*" associated with Covid-19, a respiratory disease, and the dying voice of George Floyd in 2020, which reignited the Black Lives Matter movement. Dimock brings together the arts and sciences in response to this crisis to explore precarity in its double sense of hardship and resilience. Working at the intersection of the linguistic, physiological and atmospheric, she identifies

North American indigenous resilience during the pandemic, and examines the creative response to breathing of the poet Joy Harjo, as one source of distributive, reparative agency.

Pramod K. Nayar charts the rise of planetary nuclear precarity, and the protest movement against nuclearization. According to Nayar, the various generic, media and visual cultural forms of the anti-nuclear campaigns that began in 1946 after Hiroshima were cosmopolitan in origin, scope and emphasis. Protests include "One Worldism", a form of "atomic internationalism", and responses to the Marshall Islands bomb testing, uranium mining in Australia, and fallout from Chernobyl and Fukushima. Nayar proposes an aesthetics of futuristic figuration in the form of cosmograms: the radioactive hazard symbol and the doomsday clock. Cristina M. Gámez-Fernández also examines the near annihilation of human society as imagined in a post-apocalyptic climate fiction about planetary devastation, probably caused by nuclear holocaust. This story appears in three interrelated, generically different narratives: a French graphic novel, *Le Transperceneige* (1982), its cinematic adaptation, *Snowpiercer* (2013), and the first season of a TV series, *Snowpiercer* (2020). The survivors are a class-stratified microcosm of society travelling a frozen landscape in an ever-circulating train controlled by unknown predatory forces. Gámez-Fernández asks whether technology can provide a sustainable alternative for life on the planet, and what its potential impact on human vulnerability might be.

Section 2 focuses on the ways "Revised literary genres and visual formats" address ecoprecarity and, at the same time, negotiate the limits of representation. While environmental devastation and climate change are experienced on a local level, the global and long-term environmental crisis poses a problem to the human understanding. Literature and the arts have been seen as ill-equipped to deal with the invisible global realities and incapable of departing from the limited human perspective, let alone effecting change. According to Timothy Clark, the "derangement of scales" caused by climate change are inadequately captured by the conventions of literary forms (2015, 18–19). Amitav Ghosh, in *The Great Derangement*, shares Clark's skepticism about the power of the literary imagination, arguing that the novel as a form has evaded the subject altogether (see Andersen's refutation of this claim, 2020, 14, note 16).

In environmentally oriented criticism, subject-matter has traditionally eclipsed the formal properties of texts and their potential (Caracciolo 2022, 9). In recent years, however, ecocritics have increasingly argued that it is above all form, or more specifically, "the adaptation of existing forms and the evolution of new genres or sub-genres" (Parham 2021, 16) that can "elicit affective investment in addressing ecological crisis" (14), and can do so "by sharing divergent experiences (for different people, even species) of the Anthropocene, and by reconnecting human life with exponentially vaster scales: deep history, the planet Earth, the distant future" (10). The contributions in *The Cambridge Companion to Literature and the Anthropocene* give ample evidence of such disruptive forms that serve to reconstruct our knowledge (see Parham 2021, 14), as do, for example, the "revamped genres" of the captivity narrative, the elegy and the pastoral, and tragedy that Wai Chee Dimock deals with in her study *Weak Planet* (2020). Marco Caracciolo, himself a strong advocate of the adaptation of literary form to the complexities of the new epoch of the Anthropocene, shows in *Narrating the Mesh* (2021) how his texts stage nonlinearity, interdependency and multiscalarity, while in *Contemporary Fiction and Climate Uncertainty* (2022) he focuses on "formal features that enable contemporary fictional narrative to negotiate the many unknowns of the ecological crisis" (7), among them the disruption of temporal linearity, spatial devices, metalepsis or the potentials of interactive storytelling in digital media.

The contributions to this section all centre on form and testify that literature can be what Wai Chee Dimock has called a "crisis-responsive art form" (2020, 9). Klara Machata's contribution focuses on the speculative fiction of US-based Indian writer Vandana Singh. Her exemplary reading of Singh's aptly titled novella "Entanglement" shows how the multiple interwoven plotlines serve to narrativize the interdependences and manifold relationships between humans, nonhumans and the material world, activating readers to broaden their perspective to the planetary and to ultimately envisage a sense of planetary solidarity.

Chiara Lanza focuses on petrofiction and thus the socio-environmental precariousness of lives affected by oil extraction. She reads Imbolo Mbue's novel *How Beautiful We Were* (2021) against the carbon economy's "cultures of concealment" as a text that makes the horrors of environmental degradation, alienation and dispossession tangible, and addresses at the same time alternative scenarios of hope and resistance to past, current, and future challenges to justice and equity. The contribution by Jan Rupp deals with the currently thriving genre of "performance ecopoetry" in the context of international climate action. Rupp's detailed analysis of poetic form shows that whether the poems combine and work towards creating a global public sphere to attend to the planet's environmental precarity or whether they invoke nostalgia for unspoilt nature and other (Romantic) poetic tropes, depends on their form and mediatization alone.

Not surprisingly, the Covid-19 pandemic has triggered scholarly interest in pandemic fictions as one form of global crisis discourse (Völkl, Obermayr and Hobisch 2023, 9; also see Däwes 2021; Däwes 2022; Pulugurtha 2023). The chapter by Scott Slovic analyses three contemporary novels by American authors which feature pandemics as examples of planetary and human precarity. Following Kate Rigby's (2015) argument about the capacity of pandemic novels to facilitate mindfulness and critical self-reflection, Slovic explores the ways in which these pandemic novels represent precarity not only as a negative condition, but also as bearing potential for an elevated sense of self-reflective vigilance.

Section 3 focuses on "Affective ecoprecarity: Relationality, resilience and resistance". Underpinning the contributions to this section is "[a]ffect theory [that] describes how, for example, aesthetic experiences, ideas or an environment can stimulate body-based sensations", according to John Parham (2021, 14). Parham points out the ways in which literary texts and visual art forms can employ "affective techniques such as shock, derangement or fear" (14) to elicit responses on the part of viewers and readers, down to "extractivist nostalgia" or "petromelancholia" (LeMenager 2023, 9). Sara Ahmed's (2004) notion of "affective economies" draws attention to the extent to which social hierarchies impact feelings, as well as to the fact that emotions "align individuals with communities – or bodily space with social space – through the very intensity of their attachment" (119). Affect is thus inherently relational, situated in between mind and matter, nature and culture, reader and text, emerging "in the capacities to act and be acted upon" (Seigworth and Gregg 2010, 1), which Donna Haraway (2008) has captured in the term "response-ability" (71). The relationality and indeterminacy of affect raise awareness of the entanglement of human and nonhuman agencies, it undermines our thinking in binaries, such as the objectification of nonhuman entities as resources, and, beyond human exceptionalism, demands our responsive engagement with what it means to be human in the Anthropocene and how to safeguard the habitability of our planet for all species and embodied entities. According to Marco Caracciolo, the form of narrative is always embodied and affective (2021, 4–5), and "narrative affects", writes Heather Houser (2014), "reveal the ecological systems within which we are enmeshed to be more than biological processes; they are coterminous with our own bodies and call out for aesthetic and ethical response" (30; also see Dederichs 2023, 245).

Scholars agree that, apart from historical reflection, it is the affective quality of narratives in the widest sense that fosters resistance to hegemonic humanisms, as well as resilience and critical thinking (see, e.g., Caracciolo 2022, 3), what Kate Rigby (2015) has called "learning more skilfully to 'dance'" with disaster (4). The notion of resilience, defined as the capacity of human and nonhuman beings "to adapt, renew, even thrive in the face of adversity" (Fraile-Marcos 2020, 2) has been used "to analyze […] the response of our planet's ecosystems to the impact of human action" (2). More recently, it has been criticized as a discourse of neoliberal governance which promotes individual responsibility and serves the maintenance of systemic structures, stalling agency and resistance (Fraile-Marcos 2020, 4; see also Butler and Worms 2023, 28).

Rather than drawing a line between agency and passivity or bowing before the impossibility to imagine other worlds, the contributions to this section discuss texts that focus on the embodied entanglement of human and nonhuman life forms that renders their characters precarious and at the same time recovers their agency. It is relationality, "a core concept of the environmental humanities" (Bartosch 2017, 34), and planetary interconnectedness that the texts foreground, by exploring "queernaturecultures", by envisaging a tradosphere that allows us to think backward and forward through the earth mother Gaia, and by calling our attention to the long tradition of ecopoets and ecophilosophers who view Earth as a living entity and as life's home.

Stefan Benz analyses Nicky Drayden's 2019 space opera *Escaping Exodus* with a view to its proposal of a queer ecological and queer utopian perspective as an epistemological and ethical alternative to current divisive and exploitative ideologies and practices. By staging queer human and nonhuman intimacy, the novel foregrounds intra- and interspecies interdependencies, presenting at the same time a strong critique of sexism and heteronormativity, of classism and anthropocentrism. The chapter also addresses the potential of speculative fiction in general, and queer utopias in particular to envisage strategies of survival, joint human and nonhuman agency as well as ethical and political alternatives.

In her contribution Sonja Frenzel explores the ways in which past, present and future precarity and planetary habitability are envisaged in the novels *Girl, Woman, Other* (2020) by Bernardine Evaristo and *Ada's Room* ([2020] 2023) by Sharon Dodua Otoo. Her concept of the tradosphere serves not only to relate the novels to a tradition of female writing; it also links the trope of motherhood in both novels back to earth mother Gaia, acknowledging the Earth as a life-sustaining habitat, and, translating from past experiences into future potentiality, forward to facing our complicity in ecological exploitation and our responsibility to imagine a more sustainable future. Both novels are analysed as examples of decolonizing literature in terms of gender, race, and anthropocentrism.

Leonor María Martínez Serrano focuses on the ecological thinking of Canadian poet Peter Sanger as expressed in his essay *Oikos* (2014), the title referring to the home life has built for itself and therefore synonymous in Sanger's thinking with Earth, planet, or biosphere. She traces Sanger's approach back to a centuries-old tradition of non-anthropocentric humanism that centres the biosphere, ascribes livingness and mind to it, and is alert to the vulnerability of the Earth and its life forms. Sanger requires us not only to reimagine how we as a species relate to the world but also to take seriously our responsibility to protect the Earth and its life forms.

Section 4 deliberates issues of "Planetary repair and survival", emphasizing that planetary survival requires a cognitive approach to a methodology of collective care and repair. The damage caused by planetary precarity is not limited to any particular country. Underlining the notion of the interconnectedness of planetary life, Amitav Ghosh recognizes our perilous presence, highlighting stories of ecological awareness for onward thinking about the future hazards of climate change, and claims "seeing is one of [its] central themes; *not* seeing is another" (2016, 29; original emphasis). Likewise, Bruno Latour (2018) argues that "we can understand nothing about the politics of the last 50 years if we do not put the question of climate change and its denial front and center" (2).

The discourse of planetary precarity suggests that the human world cannot be seen as isolated, hence Elizabeth DeLoughrey debunks the "myth" of isolates in the context of nuclear tests:

> The lie of isolation has indeed been a dangerous game, to the Marshall Islanders especially, and beyond. Due to these thermonuclear weapons, the entire planet is permeated with militarized radiation […]. Radioactive elements produced by these weapons were spread through the atmosphere, deposited into water supplies and soils, absorbed by plants and thus into the bone tissue of humans all over the globe. The body of every human on the planet now contains strontium90 a man-made byproduct of nuclear detonations […]. At very conservative estimates, these nuclear weapons tests have produced 400,000 cancer deaths worldwide.
>
> (2012, 179)

Peter Sloterdijk (2009) terms this irradiated violence "atmosterrorism", extending from the American West, Los Alamos, the 1945 Trinity Bomb test site, to the Pacific Islands, and at any point in time capable of annihilating or causing irreparable damage to the Earth's ecosystem.

The fact that nuclearization is an irreversible phenomenon and needs to be abandoned prompts Pramod K. Nayar to state: "the work of nuclear cultures is the prognosis of planetary precarity" (2023, 11). No wonder then that Gabriela Schwab advocates an urgent need to discuss the ramifications of "radioactive ghosts" (2020), arguing that "while biopolitics and ecology intersect in debates about climate change, the entanglement between climate change and the nuclear threat is all but eclipsed from these debates" (xi–xii). Nayar rightly suggests that such mutations of ecological disaster lead to planetary precarity. The uncertainty and damages triggered by humans' encroachment on nature and chemicalization of life have resulted in the condition of ecoprecarity, which, he points out, could be witnessed in "Katrina, Fukushima, the tsunami and, of course, Bhopal and Chernobyl" (2019, 7). Nayar expands the theoretical purchase of ecoprecarity by bringing the planet's entire ecology under its umbrella, situating the Anthropocene as its catalytic moment. Likewise Jedediah Purdy (2015) argues that the natural and the artificial have converged at every scale in the age of the Anthropocene. Purdy suggests that the crises "will require a politics that finds a way to merge, or at least hold together, certain questions that we have called ecological and others that we have called humanitarian, questions of conservation and questions of justice" (6).

In her influential book, *Weak Planet: Literature and Assisted Survival*, Wai Chee Dimock (2020) provides a phenomenological account of humans' vulnerability and insecurities and ties them to the major challenge of planetary habitability. The task of human agency, according to Dimock, therefore, should be to initiate interactions, collaborations, leading modes of resilience against oppressive structures and regimes. Dimock emphasizes the importance of this kind of relationship and calls for ways to work collectively towards making such connections, weak as they may appear, across species and geographical locations. The establishment of such connections is what Dimock identifies as "assisted survival". Dimock advocates strong networks of social assets that require "an infrastructure, a support team on permanent standby, making up for imperfect outcomes with incessant labor" (7).

Dimock's views on "assisted survival" can be seen as "an expansive notion of relatedness" (Flower and Hamington 2022, 2). Michael Flower and Maurice Hamington (2022) advocate that "inclusive connection is the empathetic and emotional attachment that we form, which motivates and participates in our engagement; and responsive action consists of the practices required to meet the needs of the other" (3). They suggest that a "plural" and "multimodal" affair with other species is potentially viable to develop an ethics of planetary care and repair (5).

Consequently, the chapters in this section deal with critical frameworks of planetary care and survival. Aleks Wansbrough's chapter analyzes Larissa Sansour's and Søren Lind's (2016) video work/film, *In the Future They Ate from the Finest Porcelain*. In referring to this film, Wansbrough defines a "planetary aesthetic", focusing on the scorched Earth after the bombing of Palestinian lands. While this ground is both literal and symbolic of bombings by the state of Israel, the chapter also conjures what Emily Apter (2013) terms "planetary dysphoria", a condition of political and ecological despair due to the limited political possibilities within the Anthropocene, and expressing the powerlessness people feel in relation to climate change. Advancing Walter Benjamin's ([1940] 1968) notion of the philosophy of history, Wansbrough argues that in recreating a collective future, it would be a mistake to approach history as a narrative – the frozen past. Instead history should be taken as an opportunity for future reconstitution. From this perspective history is understood to be related to narrative and subject to imaginative reinvention. An imaginative reach is needed to play an interventionist role in developing models of planetary care.

The following chapter by Kanak Yadav complements Wansbrough's by arguing that environmental precarity cannot be viewed as distinct from the risky situations and political instabilities that humans deal with. The chapter emphasizes that the need for planetary repair and survival has inspired a robust "zero-waste" cultural politics. Further, Yadav argues that the cultural practice of zero-waste, its digital aesthetics and the individual-centric solution create a third wave of environmentalism. However, the planetary solution offered by zero-waste politics remains "precarious" as the policy is dominated by a neoliberal agenda and as such, is limited to individual choices and eco-capitalism, and lacks any form of regulation that can slow down the ongoing environmental catastrophe.

The last chapter by Harmattan Theater founder and director May Joseph and Sofia Varino overviews the Theatre's pioneering activist "blue humanities" engagement with oceans and water. It proposes hydrosophy as an environmental methodology and aesthetic practice informed by Félix Guattari's ([1989] 2000) "ecosophy" and his concept of "three ecologies". The chapter examines site-specific performance collaborations with oceans and their multispecies communities, alongside forgotten histories and affective memories of water against landscape, the ecologies of hydrosophy, of water thinking, of water coming to engage with the challenges of rising seas and water levels. Joseph's and Varino's praxis can be read in light of Sabine Wilke's (2019) comment: "These entanglements are examples of Anthropocene interweavings that highlight the common responsibility of human culture to embrace deep history and culture as an additional perspective for analysing the meaning and functioning of individual cultures and societies, indigenous or otherwise" (144). Joseph and Varino suggest that through such performances, local species come alive in a way we can respond to and collaborate with as artists and earthlings, inhabitants of a planet on the brink of mass extinction. This responsiveness inaugurates an embodied ecological ethics as invested in political in/action as it is in experimenting with communal ties and social relations.

In conclusion, in dealing with the topic of planetary ecological precarity this collection is intended as an intervention into current postcolonial, ecocritical discourses concerned with the precariousness of life in the Anthropocene. Its arguments stem from the belief that the historic exploitation and oppression of colonization is inextricably linked to climate change, and hence present-day neocolonial and neoliberal practices actively contribute to planetary precarity. Drawing upon the philosophy of pandemics and using the response to Covid-19 as well as Dimock's notion of "assisted survival" as models for repair and recovery, the different contributors advocate an ethics of collective responsibility and care for future planetary habitability in response to human and nonhuman vulnerability. They argue that relationality and collaboration are crucial to the development of support systems, even while recognizing that "awareness of interconnectedness and coexistence" may not necessarily lead to agentive change (Woods 2018, 504).

Such a proposed reorientation of anthropogenic agency includes a decolonizing approach to western categories of thought, an embrace of alternative epistemologies and indigenous systems of healing and belief, as well as classical and modern western cultural traditions of philosophy which celebrate the Earth as a living breathing entity. It extends to collective forms of resistance such as anti-nuclear campaigns and the media of protest, as well as performances of interactive encounters between land and sea. It encompasses insight into solidarity and interconnectedness that expands knowledge into deep time and space, and resilience in order to adjust, adapt and live with the crisis as well as imagine solutions to it. In this heuristic methodology, whether any one approach will work better than any other is not possible to predict, but this collection is presented with the proviso that the proposals are of our time and belong to the search for what is possible within existing frameworks as part of the collective creative response that seeks to make planet Earth a shared home for all.

References

Ahmed, Sara. 2004. "Affective Economies." *Social Text* 22 (2): 117–139.

Andersen, Gregers. 2020. *Climate Fiction and Cultural Analysis: A New Perspective of Life in the Anthropocene*. London and New York: Routledge.

Apter, Emily. 2013. "Planetary Dysphoria." *Third Text* 27 (1): 131–140.

Armstrong McKay, David I., Arie Staal, Jesse F. Abrams, Ricarda Winkelmann, Boris Sakschewski, Sina Loriani, Ingo Fetzer, Sarah E. Cornell, Johan Rockström, and Timothy M. Lenton. 2022. "Exceeding 1.5°C Gobal Warming Could Trigger Multiple Climate Tipping Points." September 9. *Science* 377. eabn 7950. https://doi.org/10.1126/science.abn7950

Ashcroft, Bill, Gareth Griffiths, and Helen Tiffin, eds. 2006. *The Post-Colonial Studies Reader*. 2nd edition. London and New York: Routledge.

Bartosch, Roman. 2017. "Æsthetic Æffect: Relationality as a Core Concept in Environmental Studies and Education." In *Ecocriticism – Environments in Anglophone Literatures*, edited by Sonja Frenzel and Birgit Neumann, 33–57. Heidelberg: Winter.

Baxi, Upendra. 2020. "Exodus Constitutionalism: Mass Migration in Covid Lockdown Times." *The India Forum*. June 11. Accessed April 20, 2024. https://www.theindiaforum.in/article/exodus-constitutionalism

Benjamin, Walter. [1940] 1968. "Theses on the Philosophy of History." In *Illuminations*, translated by Harry Zohn, 253–264. New York: Schocken Books.

Bracke, Astrid. 2018. *Climate Crisis and the 21st-Century British Novel*. London: Bloomsbury.

Butler, Judith. 2004. *Precarious Life: The Powers of Mourning and Violence.* London and New York: Verso.
Butler, Judith. 2009. *Frames of War: When Is Life Grievable?* London and New York: Verso.
Butler, Judith. 2015a. "Foreword." In *State of Insecurity: Government of the Precarious*, edited by Isabell Lorey, vii–xi. London: Verso.
Butler, Judith. 2015b. *Notes toward a Performative Theory of Assembly.* Cambridge: Harvard University Press.
Butler, Judith. 2016. "Rethinking Vulnerability and Resistance." In *Vulnerability in Resistance*, edited by Judith Butler, Zeynep Gambetti, and Leticia Sabsay, 12–27. Durham, NC: Duke University Press.
Butler, Judith, and Frédéric Worms. 2023. *The Livable and the Unlivable.* A Conversation initiated by Arto Charpentier and Laure Barillas. Translated by Zakiya Hanafi. New York: Fordham University Press.
Caracciolo, Marco. 2021. *Narrating the Mesh: Form and Story in the Anthropocene.* Charlottesville, VA: University of Virginia Press.
Caracciolo, Marco. 2022. *Contemporary Fiction and Climate Uncertainty: Narrating Unstable Futures.* London: Bloomsbury.
Carrigan Anthony, 2011. *Postcolonial Tourism: Literature, Culture, and Environment.* New York: Routledge.
Chakrabarty, Dipesh. 2021. *The Climate of History in a Planetary Age.* Chicago, IL and London: University of Chicago Press.
Clark, Timothy. 2015. *Ecocriticism on the Edge: The Anthropocene as a Threshold Concept.* London: Bloomsbury.
Cooke, Stuart, and Peter Denney. 2021. "Thinking About Transcultural Ecocriticism: Space, Scale and Translation." In *Transcultural Ecocriticism: Global, Romantic and Decolonial Perspectives*, edited by Stuart Cooke and Peter Denney, 1–20. London: Bloomsbury.
Crutzen, Paul J., and Eugene F. Stoermer. 2000. "The Anthropocene." *The International Geosphere-Biosphere Programme Newsletter* 41: 17–18.
Danewid, Ida. 2017. "White Innocence in the Black Mediterranean: Hospitality and the Erasure of History." *Third World Quarterly* 38 (7): 1674–1689.
Däwes, Birgit. 2021. "'Untenanted by any Tangible Form': Illness, Minorities, and Narrative Masquerades in Contemporary Pandemic Fiction." *On Culture: The Open Journal for the Study of Culture* 11: 1–28. https://www.on-culture.org/journal/issue-11/untenanted-by-any-tangible-form/
Däwes, Birgit. 2022. "Molecular Mimicry, Realism, and the Collective Memory of Pandemics: Narrative Strategies of COVID-19 Fiction." *DIEGESIS. Interdisciplinary E-Journal for Narrative Research/E-Journal für Erzählforschung* 11 (1): 1–24. https://www.diegesis.uni-wuppertal.de/index.php/diegesis/article/download/432/617
Dederichs, Natalie. 2023. *Atmosfears: The Uncanny Climate of Contemporary Ecofiction.* Bielefeld: transcript.
DeLoughrey, Elizabeth. 2012. "The Myth of Isolates: Ecosystem Ecologies in the Nuclear Pacific." *Cultural Geographies* 20 (2): 167–184.
Dimock, Wai Chee. 2020. *Weak Planet: Literature and Assisted Survival.* Chicago, IL: University of Chicago Press.

During, Simon. 2015. "From the Subaltern to the Precariat." *Boundary2* 42 (2): 57–82. https://read.dukeupress.edu/boundary-2/article-abstract/42/2/57/56103/From-the-Subaltern-to-the-Precariat?redirectedFrom=fulltext.

Dwivedi, Om Prakash. 2023. "Subaltern Ecologies: Cultures of Concealment and Carbon Economy in Amitav Ghosh's *The Great Derangement*." *Postcolonial Text* 18 (1&2): 1–16.

Fineman, Martha Albertson. 2021. "Populations, Pandemics and Politics." In *Lessons for and From Vulnerability Theory*. Special Issue of *International Journal of Discrimination and the Law* 21 (3): 184–190. https://journals.sagepub.com/doi/10.1177/13582291211042212. https://doi.org/10.1177/13582291211042212

Flower, Michael, and Maurice Hamington. 2022. "Care Ethics, Bruno Latour, and the Anthropocene." In "Feminist Care Ethics Confronts Mainstream Philosophy", edited by Maurice Hamington and Maggie FitzGerald, special issue of *Philosophies* 7 (2): 1–18. https://doi.org/10.3390/philosophies7020031

Fraile-Marcos, Ana María. 2020. "Introduction: Glocal Narratives of Resilience and Healing." In *Glocal Narratives of Resilience*, edited by Ana María Fraile-Marcos, 1–20. New York and London: Routledge.

Gámez-Fernández, Cristina M., and Miriam Fernández-Santiago. 2023. "An Ecology of Embodied VulnerAbility—Literary and Filmic Representations across the Globe." In *Embodied VulnerAbilities in Literature and Film*, edited by Cristina Gámez-Fernández and Miriam Fernández-Santiago, 1–16. London: Routledge.

Ghosh, Amitav. 2016. *The Great Derangement: Climate Change and the Unthinkable*. State: Chicago, IL: University of Chicago Press.

Guattari, Félix. 2000. *The Three Ecologies*. Translated by Ian Pindar and Paul Sutton. London and New Brunswick: The Athlone Press.

Haraway, Donna J. 2008. *When Species Meet*. Minneapolis, MN: Minnesota University Press.

Hochberg, Gil. 2018. "'Jerusalem, We Have a Problem': Larissa Sansour's Sci-Fi Trilogy and the Impetus of Dystopic Imagination." *The Arab Studies Journal* 26 (1) (Spring): 34–57.

Hofmeyr, Isabel. 2019. "Provisional Notes on Hydrocolonialism." *English Language Notes* 57 (1): 11–20.

Höglund, Johan. 2020. "Challenging ecoprecarity in Paolo Bacigalupi's Ship Breaker trilogy." In "Challenging Precarity", edited by Janet M. Wilson, Om Prakash Dwivedi, and Cristina M. Gámez-Fernández, special issue, *Journal of Postcolonial Writing* 56 (4): 447–459.

Houser, Heather. 2014. *Ecosickness in Contemporary U.S. Fiction: Environment and Affect*. New York: Columbia University Press.

Huggan, Graham, and Helen Tiffin. [2010] 2015. *Postcolonial Ecocriticism: Literature, Animals, Environment*. London and New York: Routledge.

Jepson, Paul, and Cain Blythe. 2020. *Rewilding: The Radical New Science of Ecological Recovery*. London: Icon Books Ltd.

Kerridge, Richard, and Neil Sammells. 1998. *Writing the Environment: Ecocriticism and Literature*. London: Zed Books.

Latour, Bruno. 2018. *Down to Earth: Politics in the New Climatic Regime*. Cambridge: Polity Press.

LeMenager, Stephanie. 2023. “Preface.” In *To the Last Drop: Affective Economies of Extraction and Sentimentality*, edited by Axelle Germanaz, Danielea Gutiérrez Fuentes, Sarah Marak, and Heike Paul, 7–11. Bielefeld: transcript.

Lenton, Timothy M, David I. Armstrong McKay: Sina Loriani, Jesse F. Abrams, Steven J. Lade. Jonathan F. Donges, Manjana Milkoreit, Tom Powell, Steven R. Smith, Caroline Zimm, Emma Bailey, Joshua E. Buxton, James G. Dyke, Ashish Ghadiali, snd Laurie Laybourn. eds. 2023. *The Global Tipping Points Report 2023*. University of Exeter, UK. https://report-2023.global-tipping-points.org/

Lesutis, Gediminas. 2022. *The Politics of Precarity: Spaces of Extractivism, Violence, and Suffering*. London and New York: Routledge.

Martin, Julia. 2017. “Review of *Africa. Ecocriticism and Animal Studies in Contemporary Cultural Forms*, edited by Fiona Muller.” *Tydskrif vir Letterkunde* 54 (2): 177–180. https://letterkunde.africa/article/view/2979; https://doi.org/10.17159/2309-9070/tvl.v.54i2.2979.

Mbembe, Achille. 2019. “Thoughts on the Planetary.” *New Frame*. September 5. https://www.newframe.com/thoughts-on-the-planetary-an-interview-with-achille-mbembe/

McGrath, Matt, and Mark Poynting. 2024. “Climate Change: ‘Uncharted Territory’ Fears After Record Hot March.” *BBC News Climate & Science*, April 9. https://www.bbc.co.uk/news/science-environment-68665166

Michaels, Peter. 2021. “Love and Social Distancing in the Time of Covid-19: The Philosophy and Literature of Pandemics.” *Educational Philosophy and Theory* 53 (8): 755–759.

Mišík, Matúš, and Nada Kujundžić, eds. 2021. *Energy Humanities: Current State and Future Directions*. Cham: Springer.

Moolla, Fiona F. 2016. “Introduction.” In *Natures of Africa. Ecocriticism and Animal Studies in Contemporary Cultural Forms,* edited by F. Fiona Moolla, 1–26. Johannesburg: Wits University Press.

Moore, Jason W. 2015. *Capitalism and the Web of Life: Ecology and the Accumulation of Capital.* London and New York: Verso.

Moore, Jason W. 2016. “Introduction. Anthropocene or Capitalocene? The Nature, History and Crisis of Capitalism.” In *Anthropocene or Capitalocene? Nature, History, and the Crisis of Capitalism*, edited by Jason W. Moore, 1–11. Oakland, CA: PM Press.

Nayar, Pramod K. 2019. *Ecoprecarity: Vulnerable Lives in Literature and Culture.* London and New York: Routledge.

Nayar, Pramod K. 2023. *Nuclear Cultures: Irradiated Subjects, Aesthetics and Planetary Precarity*. New York and London: Routledge.

Nayar, Pramod K. 2024. *Vulnerable Earth: The Literature of Climate Crisis.* Cambridge: Cambridge University Press.

Nixon, Rob. 2011. *Slow Violence and the Environmentalism of the Poor.* Cambridge, MA: Harvard University Press.

Nixon Rob. 2014. “The Great Acceleration and the Great Divergence: Vulnerability in the Anthropocene.” *Profession* 19. https://profession.mla.org/the-great-acceleration-and-the-great-divergence-vulnerability-in-the-anthropocene/

Oldfield, Frank, Anthony D. Barnosky, John Dearing, Marina Fischer-Kowalski, John McNeill, Will Steffen, and Jan Zalasiewicz. 2014. “*The Anthropocene*

Review: Its Significance, Implications and the Rationale for a New Transdisciplinary Journal." *The Anthropocene Review* 1 (1): 3–7.
Otto, Ilona M., Jonathan F. Donges, Roger Cremades, Avit Bhowmik, Richard J. Hewitt, Wolfgang Lucht, Johan Rockström, Franziska Allerberger, Mark McCaffrey, Sylvanus S.P. Doe, Alex Lenferna, Nerea Morán, Detlef P. van Vuuren, and Hans Joachim Schellnhuber 2020. "Social Tipping Dynamics for Stabilizing Earth's Climate by 2050." February 4. *Science* 117 (5): 2354–2365. https://doi.org/10.1126/science.abn7950
Parham, John. 2021. "Introduction – With or Without Us: Literature and the Anthropocene." In *The Cambridge Companion to Literature and the Anthropocene*, edited by John Parham, 1–33. Cambridge: Cambridge University Press.
Parham, John. 2023. "Introduction". In *The Literature and Politics of the Environment*, edited by John Parham, 1-5. Cambridge: D.S. Brewer.
Poray-Wybranowska, Justyna. 2021. *Climate Change, Ecological Catastrophe, and the Contemporary Postcolonial Novel*. London and New York: Routledge.
Pulugurtha, Nishi, ed. 2023. *Literary Representations of Pandemics, Epidemics and Pestilence*. London and New York: Routledge.
Purdy, Jedediah. 2015. *After Nature: A Politics for the Anthropocene*. Cambridge, MA: Harvard University Press.
Rigby, Kate. 2015. *Dancing with Disaster: Environmental Histories, Narratives, and Ethics for Perilous Times*. Charlottesville, VA: University of Virginia Press.
Rose, Deborah Bird, and Thom Van Dooren. 2017. "Encountering a More-than-human World: Ethos and the Arts of Witness." In *The Routledge Companion to the Environmental Humanities*, edited by Ursula Heise, Jon Christensen, and Michelle Niemann, 120–128. London: Routledge.
Sabsay, Leticia. 2016. "Permeable Bodies: Vulnerability, Affective Powers, Hegemony." In *Vulnerability in Resistance*, edited by Judith Butler, Zeynep Gambetti, and Leticia Sabsay, 278–302. Durham, NC: Duke University Press.
Schwab, Gabriele. 2020. *Radioactive Ghosts*. Minneapolis, MN: University of Minnesota Press.
Schmidt-Haberkamp, Barbara, and Marion Gymnich, 2022. "Representing Poverty and Precarity in a Postcolonial World: An Introduction." In *Representing Poverty and Precarity in a Postcolonial World*, edited by Barbara Schmidt-Haberkamp, Marion Gymnich and Klaus P. Schneider, 1–18. Leiden and Boston, MA: Brill.
Seigworth, Gregory J., and Melissa Gregg. 2010. "An Inventory of Shimmers." In *The Affect Theory Reader*, edited by Melissa Gregg and Gregory J. Seigworth, 1–25. Durham, NC and London: Duke University Press.
Sloterdijk, Peter. 2009. "Airquakes." [Excerpt from Vol. 3 of *Sphären* (Spheres). "Schäume." (Foam)]. Translated by Eduardo Mendieta. *Environment and Planning D: Society and Space* 27 (1): 41–57.
Slovic, Scott. 2020. "COVID World, COVID Mind: Toward a New Consciousness." In "The New Normal? An Environmental Humanities Response", edited by Steven Hartman, Joni Adamson, Greta Gaard, and Serpil Oppermann, June 8, special issue, *Bifrost Online* https://bifrostonline.org/scott-slovic/
Standing, Guy. 2011. *The Precariat: The New Dangerous Class*. London: Bloomsbury Academic.

Szeman, Imre, and Dominic Boyer, eds. 2017. *Energy Humanities: An Anthology*. Baltimore, MD: Johns Hopkins University Press.

Thobani, Sunera. 2010. "White Innocence, Western Supremacy: The Role of Western Feminism in the War on Terror." In *States of Race: Critical Race Feminism for the 21st Century*, edited by Sherene Razack, Malinda Smith, and Sunera Thobani, 127–146. Toronto: Between the Lines.

Trexler, Adam. 2015. *Anthropocene Fictions: The Novel in a Time of Climate Change*. Charlottesville, VA: University of Virginia Press.

Tsing, Anna Lowenhaupt. 2015. *The Mushroom at the End of the World: On the Possibility of Life in Capitalist Ruins*. Princeton, NJ: Princeton University Press.

Völkl, Yvonne, Julia Obermayr, and Elisabeth Hobisch. 2023. "Pandemic Protagonists (Re)Claiming Agency: An Introduction." In *Pandemic Protagonists: Viral (Re)Actions in Pandemic and Corona Fictions*, edited by Yvonne Völkl, Julia Obermayr, and Elisabeth Hobisch, 9–22. Bielefeld: transcript.

Wallace, Molly. 2017. "It's the End of the Field as We Know It (and I Feel Fine)." *American Literary History* 29 (3): 565–578.

Wilke, Sabine. 2019. "Theatre and Performance." In *Routledge Handbook of Diaspora Studies*, edited by Robin Cohen and Carolin Fischer, 132–146. London: Routledge.

Wilson, Janet M., Om Prakash Dwivedi, and Cristina M. Gámez-Fernández. 2020. "Planetary Precarity and the Pandemic." Introduction to "Challenging Precarity", edited by Janet M. Wilson, Om Prakash Dwivedi, and Cristina M. Gámez-Fernández, special issue, *Journal of Postcolonial Writing* 56 (4): 439–446.

Woods, Derek. 2018. "*Ecocriticism on the Edge: The Anthropocene as a Threshold Concept* by Timothy Clark." *Configurations* 26 (4): 502–504.

Part I

Planetary precarity and vulnerability

1 Precarious breath

The arts and sciences of oxygen

Wai Chee Dimock

Introduction: oxygen, the breath of life

I can't breathe.[1] It began as a medical symptom, an oppressive blockage in the respiratory system that put the word "ventilator" on everyone's lips. Then it was a cry of desperation: the last words from Eric Garner back in 2014, since then repeated by 70 others, and most agonizingly by George Floyd.[2] And finally, in an outsized sequel, the words have become elemental, apocalyptic, with raging wildfires and plumes of orange smoke engulfing the American West, pollution so massive as to be felt on the East Coast and even Europe.

It is not entirely accidental that the three crises should converge in a single year. Nor is it accidental that all three should rest on a simple physiological fact: our need to breathe, to have free and unrestricted access to oxygen. Oxygen is the one resource on the planet we have always counted on. It gives us life even as it connects us to other living things. Keeping the oxygen flowing is not always easy, because humans happen to have a respiratory system that is highly evolved but also delicate and vulnerable. It can be easily sabotaged, putting us immediately at risk, in danger of becoming oxygen-deprived on any number of fronts.

Even at rest, the average human breathes 10–20 times per minute. Between 2,100 and 2,400 gallons of air pass through our lungs every day. This nonstop oxygen intake exposes us to a full range of pathogens and pollutants in the environment. Covid-19 is a respiratory disease, so too are many other new infectious diseases, like SARS and MERS. As for wildfire smoke, we know now that it is rich in toxins called PM2.5 – articulate matter of less than 2.5 microns – that can travel thousands of miles and damage our immune system weeks and even months into the future (see Herschlag 2020, n.p; Gibbens and McKeever 2020, n.p.). In the 21st century, every breath we take is a risk taken, a risk shared with every other creature on the planet.

DOI: 10.4324/9781003383598-3

Precarity, Covid-19, and oxygen shortages

Precarity, seen through the lens of oxygen, is at once monstrously tangible and complexly diffused, with many of its causes coming from unknown locales in the past, and many of its downstream effects not materializing until some distance into the future. Not limited to a single outcome, it is also not analyzable by a single metric. Evolving across space and time, it is shaped at every turn by forces both large and small, some of them human and some not.

Especially at this juncture, with climate change looming ever larger on the horizon, and COVID variants still ravaging the globe, it is important to think about precarity as a human-nonhuman interface, the combined effect of chemistry, biology, and politics. Oxygen is a useful index here, taking us from the physiology we inhabit to the atmosphere we share, to the public health infrastructures unique to particular jurisdictions. From winter 2020 well into summer 2021, oxygen was in short supply across the globe, arising from different circumstances but resting on the same biological ground. This shared lack suggests that precarity, though experientially devastating, could nonetheless be an important unifying force, the beginning of a reparative momentum cutting across political, economic, and religious divides, and giving our creative energies salient outlets.

India is a case in point. In April and May 2021, severe oxygen shortage, coupled with lack of ICU beds, left patients dying in ambulances and parking lots, while family members stood in queues for hours hoping to fill their oxygen cylinders. Public health in India has long been underfunded, amounting to just 1 percent of its gross domestic product, compared with 18 percent in the US and 9 percent in Brazil. With bureaucratic bottlenecks and no coordinated transport, oxygen remained in the east and south of the country, where it was produced, useless to those who desperately needed it in the west and north (Frayer 2021, n.p.).

In the midst of this crisis, Sikh organizations like Hemkunt Foundation and Khalsa Aid stepped up to help. Sikh temples have always provided "langars" (free meals) for migrant workers and others in need. Earlier in 2021, they had got into trouble when they expressed solidarity with the hundreds of thousands of farmers protesting against Narendra Modi's agricultural laws. Modi's supporters had branded them "Khalistani", or secessionist. That didn't stop them from setting up "oxygen langars" now, backed by a volunteer delivery system and the belief that every human life deserves respect because of the indwelling *Waheguru*, or divine spark (Chabba 2021, n.p.).

United Sikhs, an UN-affiliated human rights organization, soon joined them. Opening up COVID care centers across India, this tech-savvy NGO made creative use of media platforms to connect with those who needed oxygen (Anon 2021, n.p.). Unexpected financial support in turn came through these channels. Lil Ray Ray, a six-year-old Instagram icon, launched an urgent fund-drive among his many followers, with impressive results. United Sikhs ended up receiving a check for £20,867 to fund its shipment of oxygen concentrators, cylinders, and oximeters from the US, Canada, and Malaysia. As with the other Sikh organizations, the guiding principle is this: *Manas kee jaat ek pehchaanbo.* (We recognize the human race as one) (Singh 2021, n.p.). Mindful only of the fact that oxygen is a necessity of life, that its absence is felt in exactly the same way regardless of religious affiliation, these Sikh groups turn our precarity into a common denominator, uniting us as biological beings and sustaining us at the human/nonhuman interface.

The word "anthropogenic" becomes especially interesting in this context. Usually associated only with human-caused harm, it is no less meaningful in the context of harm partly alleviated. In what follows, I explore these contrasting senses of the word through the contrasting trajectories of oxygen. Taking these to be intertwined, but differentially expressed across the linguistic, atmospheric, and physiological spectrums, I use this multi-path, multi-scale landscape to redraw the map of the humanities, bringing together disciplines usually kept apart. Native American arts and sciences offer an especially important archive for this approach. Most often discussed only in terms of injury sustained, this body of evidence in fact also shows injury partly redressed. Speaking to both senses of the anthropogenic, it traces agency sometimes harm-bearing and sometimes harm-mitigating, suggesting that reparative agency is possible, and that diversifying our investigative lens might be one way to begin.

Covid-19 and Native Americans

When the Cherokee Nation got its first shipment of the Pfizer vaccine in December 2020, there was no question where it would go. "We put Cherokee-fluent speakers at the front of the line", said Principal Chief Chuck Hoskin Jr. "Saving the language" is a top priority at all times, and never more so than during a pandemic (Brown 2021, n.p.). Cherokee, like many other indigenous languages around the world, is endangered, with just 2,000 fluent speakers, most of them elders, 45 of whom were lost to Covid-19 (Kaur 2021, n.p.). The deaths of these keepers of language meant lost libraries of "lifeways, culture, stories". Stung by these losses, the tribe pulled together to get the vaccine out as soon as it arrived. "Our doses have been administered without any lag time", Hoskin said (Brown 2021, n.p.).

The Cherokee Nation, forced at gun point to relocate from the southeast in 1838, now has 141,000 citizens on its reservation in Oklahoma. By February 8, 2021, more than 17,000 had had both shots, a vaccination rate that, at 12 percent, was significantly higher than that of most states. Cherokee speakers, elders, and healthcare workers had all been vaccinated, and the tribe was able to open up the distribution to those 55 and older, as well as teachers and tribal government employees. On February 25, the Cherokee Nation announced those 16 and older were now eligible (Mudd 2021, n.p.). By March 11, the tribe began offering the vaccine to the general public, non-Native as well as Native (King 2021, n.p.).

Hoskin attributed this public health success to the decision to put Cherokee speakers at the head of the line. Given the festering mistrust among tribes – especially acute after the federal government sterilized about a quarter of Native women of childbearing age between 1970 and 1976 (Theobald 2019, n.p.) – nothing cleared the air better than a vaccine-protected Cherokee language. It was the "biggest confidence builder" (quoted in Kaur 2021, n.p.) anyone could have come up with. By vaccinating its revered speakers, the tribe was sending a signal no one could fail to notice. There are no anti-vaxxers in the community, because language preservation has unified the Cherokee, creating a public health mandate as well as a "sense of optimism among our people", Hoskin told CNN (quoted in Kaur 2021, n.p.).

It also helped that the infrastructure was already in place for the vaccine rollout. The Cherokee is clear-eyed about the nation's healthcare system and has acted accordingly. "We're a country that doesn't make access to health care just part of being a citizen", Hoskin said. "That's a problem during good times, and it's certainly a problem during the pandemic" (quoted in Kaur 2021, n.p.). Not leaving everything to the state of Oklahoma, or the federal government, the Cherokee Nation has been investing for years in its state-of-the-art hospital and clinics, now the largest tribally operated health system in the US. It has made headline news before. Back in November, PBS *Newshour* had a story, "How the Cherokee Nation has curtailed the Pandemic" (McFarling 2020), featuring the tribe's cutting-edge contact tracing.

The Cherokee is less of an outlier than we might think. Other tribes, without such advanced medical infrastructure, are also doing fine with the vaccine rollout, thanks to the clearly stated goal of keeping languages alive. According to an NPR analysis, Native communities as a rule have been faster than the rest of the country in getting shots into people's arms (Siegler 2021, n.p.). Given how dire things were at the beginning, these public health successes are unexpected, to say the least.

Hand-washing is difficult on reservations because Native households are 19 times as likely as white households to lack access to indoor plumbing. "Running water, it would be such a luxury", said Legena Wagner, a member of Navajo Nation (Stead Sellers 2019, n.p.). Other preexisting conditions are just as crushing. Native Americans are eight times more likely to die of tuberculosis (Centers for Disease Prevention and Control) and 2.3 times more likely to die of diabetes (US Department of Health and Human Services). During the early months of the pandemic, they were dying at twice the rate as white Americans (Currizzo 2020, n.p.). And yet, as Kyle Powys Whyte (2017, 2018) reminds us, existential threats are nothing new for Native communities, who have always found a way to come out at the other end. "Our genetic memory of past epidemics is very strong", said Chuck Sams, Covid-19 incident commander on the Umatilla Reservation: "Epidemics that came through between 1780 to 1860 wiped out nearly 95% of our population. We're the descendants of the 5% that survived" (quoted in Brown 2021, n.p.).

The Rosebud Sioux and Oglala Sioux tribes in South Dakota, with the lowest life expectancy in the US, represent that 5 percent. Among the hardest hit in the pandemic, they have nonetheless been able to bounce back by making language preservation a driving force in public health. Like Cherokee, Lakota is endangered, again with just 2,000 speakers. Elders fluent in that language offer vaccine information on the radio. Call center staffs likewise make appointments and schedule transportation in Lakota. "That's one of the defining features of our response. Our community figures things out fairly well with limited resources", said Alicia Mousseau, vice-president of the Oglala Sioux Tribe (Brown 2021, n.p.). Out of a population of about 30,000 on its reservation, the tribe had vaccinated more than 4,000 people by February 1 (Kaur 2021, n.p.). The Rosebud Sioux Tribe, meanwhile, was vaccinating at double the rate of South Dakota. "We've even had non-Indian people from Sioux Falls and Omaha trying to get in here to get vaccinated because they can't get it over there", said Rodney Bordeaux, the tribe's president (Siegler 2021, n.p.).

The Navajo Nation has no such extensive partnerships, but language as oxygen for life works here as well, in spite of past traumas. The largest reservation in the US, it has also witnessed some of the worst health crises. The three waves of the 1918–1919 flu epidemic killed about 10 percent of the Navajo population (Brady and Bahr 2014, n.p.). Covid-19 (*Dikos Nitsaa'igii-19*, "the big cough" in Navajo) has also unfolded swiftly and relentlessly. By April 4, 2020, 321 tribal members had already been infected, with an increase of 51 cases in a single day with 13 deaths (Hedgpeth et al. 2020, n.p.). By May 18, the Navajo Nation surpassed New York State with the highest infection rate in the country (Silverman et al. 2020, n.p.). That infection rate soared to 37.8 percent on November 20 (Begay 2021, n.p.).

But other trends were also beginning to emerge. While the Navajo (Diné) language, with 170,000 speakers, is not under immediate threat of extinction, reservation health workers have nonetheless been able to use it as a hook for the vaccine rollout. Daily radio call-in shows and biweekly town hall meetings in Navajo were organized to educate the public. "Utilizing our way of life and teaching helps our people feel it's okay to take the shots", said Jonathan Nez, president of the Navajo Nation (Kaur 2021, n.p.). When the first shipment of vaccines arrived in December, three out of four of the residents welcomed them with open arms. "The Navajo Nation is beating out every state's coronavirus vaccine rollout", *Forbes* reported on April 4, 2021 (Porterfield 2021, n.p.). By April 23, 87.8 percent of the eligible population had received at least one dose, and 38.4 percent had been fully vaccinated, vastly outpacing the rest of the nation (Treisman 2021, n.p.).

Here too, Native medicine is a holistic medicine, grafting Navajo traditional knowledge onto contemporary drug-based treatment. Michael Tutt, chief medical officer of the Tséhootsooí Medical Center in Fort Defiance, Arizona, cares for a population of 30,000 by integrating the art of listening.

> My grandmother told me listening is medicine. And what you say, what comes out of your mouth, is medicine. Probably stronger than what you give as medication. When you only treat patients physically, you just throw medications at them. But the Navajo way, we treat them physically, mentally, spiritually.
>
> (Flaherty 2019, n.p.)

There's a difference between medicine and medication. The former is much broader, more democratic, an art as well as a science. The Navajo word for this holistic medicine is *Hózhó*, a balanced way of walking the earth. It means that the patient has as much input into the healing process as the physician, not just being treated with pills but also being listened to. In the context of Covid-19, with the nation deeply divided, and vaccine hesitancy remaining widespread, *Hózhó* is never more urgent. Multiple studies have shown that those who are vaccine-hesitant are more likely to change their minds when their fears and anxieties are carefully listened to and discussed by those they trust (see Keith 2021, n.p.; Howley 2021, n.p.).

To put *Hózhó* into practice, Tutt brings a traditional healer to his mobile rheumatology clinic:

> Sometimes my patients are looking forward more to the traditional healer than to me. That's what I want. Because they feel good when they get a prayer, that things will be all right. To me it's a pure way of praying for somebody. It's Navajo, it's tied to the ground, tied to our surroundings, non-denominational, non-judging. Our way of praying has been going on for 500 years. And it will still be here 500 years from now.
>
> (Flaherty 2019, n.p.)

To make sure that the Navajo prayer would last 500 years, Tutt looks to the next generation to take *Hózhó* forward. The Tséhootsooí Medical Center has an adolescent care unit, Arizona's first inpatient psychiatric clinic for young Native Americans. "A lot of these kids – I call them scrappers because somehow they've survived this long – have lost their parents to drugs and alcohol", Tutt said. "We take care of them with their addiction; we give them tools to survive their daily life. But one of the things that I think makes a big difference is we instill within the program traditional teaching" (Flaherty 2019, n.p.). The Navajo Nation can have a future only if there are youngsters to carry on its way of life. These youngsters stand a chance of surviving only if they are given metaphoric oxygen – listened to, cared for, and equipped with tools, integrating the arts and sciences as life-saving knowledge now needs to be.

Joy Harjo: science, poetry, and traditional knowledge

Saving the young is indeed no easy task. Joy Harjo (2002a), a member of the Muscogee (Creek) Nation, and now serving her third term as US poet laureate, recalls the time when she was a "broke Indian student with small children" at the University of New Mexico, trying to become a painter, and so close to the edge that "my car kept trying to drive itself off the road" (xix). Friends made her go see "an older Indian woman who was psychic; they told me she helped the police find missing people". Even as the two spoke, Harjo was "counting steps and giving my body instructions on how to move so that I would not whirl and fall apart". The woman "opened my palm then quickly closed it. She saw it all, and I saw she see it all. She told me to be careful, to take care of myself". It was at this desperate juncture, this involuntary exposure to an all-knowing stranger, that "poetry showed up" (xix).

Traditional knowledge "knows" Harjo in a way different from university knowledge. She has been a missing person; now she has been found. Not something to be proud of, but it does bring a "chaos of raw inverted power" (Harjo 2002a, xix). It turns her inside out and gives her an airing. It is a way of giving birth, and "worked exactly as the process of writing a poem. It started from the inside out, then turned back in to complete a movement" (xix). Language is literally the medium of delivery here, releasing Harjo from her suicidal past and into a new and potentially life-giving future. She would come back to this idea later, in a poem written for Alva Benson, a Navajo-speaking activist, linking the breath of the newborn to the sound of a language reproducing itself:

> And the ground spoke when she was born.
> Her mother heard it. In Navajo she answered

As she squatted down against the earth
to give birth. It was now when it happened,
now giving birth to itself again and again
between the legs of women.
(Harjo 2008, 8)

In this opening stanza, the breath of life is elemental, primordial. The mother is resting on bare earth, hearing a sound coming directly to her, untouched by any form of modern mediation, and eliciting from her a response also unmediated. But this idealized account of Navajo birth isn't quite the whole story. Babies aren't usually born this way in the 21st century. In the second stanza of the poem, Harjo turns to an alternative scenario, far less idealized, but more likely to be true:

Or maybe it was the Indian Hospital
in Gallup. The ground still spoke beneath
mortar and concrete. She strained against the
metal stirrups, and they tied her hands down
because she still spoke with them when they
muffled her screams. But her body went on
talking and the child was born into their
hands, and the child learned to speak
both voices.
(2008, 8)

In this second scenario, the newborn comes into the world in a hospital, the Gallup Indian Medical Center run by the Indian Health Services, perhaps modern medicine at its most alienating. The mother is forcibly restrained. Still, the ground speaks as before, its voice coming through the concrete and mortar, and the mother still responds. This time, though, she is also screaming at the hospital staff as they gag her. The linguistic field here is as violent as can be, its bilingualism an effect of degradation and coercion. But bilingualism even of such provenance has nontrivial benefits. It gives the Navajo Nation another lifeline to the future as the child "grew up talking in Navajo, in English" (Harjo 2008, 8).

The poem is a tribute to the bilingual Alva Benson, but it is also a semi-autobiographical reflection on the tangled paths of Harjo's own life, bilingual in another sense. Born Joy Foster to a Muscogee father and a mixed-race mother of Cherokee, French, and Irish descent, Harjo took her grandmother's name when she became a member of the Muscogee (Creek) Nation. At age 16, she attended the Institute of American Indian Arts, a fine arts school that "sits on the homelands of Pueblo nations", where "the spirit and vision of Native American and Alaska Native people are the first priority" and where "traditions are rediscovered, explored, and deepened" (Institute of American Indian Arts, n.p.). When she was able to scrape together enough money to attend the University of New Mexico, however, she chose to become a pre-med. She then switched back to art, and then to creative writing in her final year, before moving on to a master's degree from the Iowa Writers' Workshop.

In her Introduction to *When the Light of the World Was Subdued, Our Songs Came Through* (2020), an anthology of Native literature, Harjo shows that her early science education has not been wasted. "Our bodies", she writes, "carry traces of where we were born and raised: oxygen isotopes from the water we drank as children are stored within the buds of our teeth" (1). Even though Harjo did not press on with her medical career, she has not forgotten her college chemistry. Isotopes are two or more forms of the same element that have different atomic masses and different physical properties due to the different number of neutrons in the nucleus, but with the same number of protons and the same position on the periodic table. Oxygen has three isotopes. Their ratio determines both the paleoclimatology of oceans and the nature of the drinking water that goes into the making of tooth enamel (see Riebeek 2005, n.p.; Pellegrini et al. 2016, 1).

Harjo is not flaunting her specialized knowledge. She is making a more basic point that goes beyond specialized knowledge, namely that chemistry shapes us as much as culture, producing variation that is unlikely to go away, unlikely to be fully reconciled. But there is no reason why variation should result in bloodshed. The three oxygen isotopes show that variation rests on a deeper common ground: all three belong to the same element, and all three contribute to the physiological processes at work in every one of us, manifesting themselves in the same sound of "Ah, Ah":

> Ah, ah cries the crow arching toward the heavy sky over the marina.
> Lands on the crown of the palm tree.
> Ah, ah slaps the urgent cove of ocean swimming through the slips.
> We carry canoes to the edge of the salt.
> Ah, ah groans the crew with the weight, the winds cutting skin.
> We claim our seats. Pelicans perch in the draft for fish,
> Ah, ah beats our lungs and we are racing into the waves.
> Though there are worlds below us and above us, we are straight ahead.
> Ah, ah tattoos the engines of your plane against the sky – away from these waters.
> Each paddle stroke follows the curve from reach to loss.
>
> (Harjo 2002b, 185)

Human differences here are embedded within a sonic continuum deeper and more various. The "you" who flies away in a plane has something in common with the crow that flies over the marina, the ocean that slaps the shore, the crew that speeds the canoe, and the pelican that waits in the wings for fish. The five are entirely distinct. They have no common identity, no common purpose. Yet, the breath of life is shared among these different life forms, and "tattoos" each of them with the sound of Ah, Ah.

Harjo, who probably has the most famous tattoo among US authors, speaks often about the glyphs on her right hand (Figure 1.1). Glyphs are an important part of the complex mediascape of indigenous peoples, reminding us that the alphabetic script is only one signifying language among others, and that languages that look different are, nonetheless, linked in kinship (see Cohen 2009; Rasmussen 2012; Glover 2014; Cohen and Glover 2014; Garcia 2020). In this case, the glyphs are of Marquesan design, done by the Tahitian tattoo artist Roonui. "The part inside my wrist, close to my heart, resembles ancestral designs of my tribal people" (Harjo 2012, n.p.), Harjo writes on her blog. And she quotes Roonui: "Polynesian tattooing is not a simple exercise in aesthetics. Polynesians carve into their body the symbols of their actions (past present or future)" (2012, n.p.). For Harjo, the glyphs are a symbol of the diverse mediums of expression she has availed herself of, and the different kinds of support that come with each

Figure 1.1 Joy Harjo and tattoo.

platform: "I use my hands for music, writing, and everything else I do. The tattoo reminds me of the levels of assistance" (n.p.).

Conclusion: giving life (to the inanimate)

A musician as well as a poet, Harjo has toured the world with her band, the Arrow Dynamics, singing, and playing the flute and alto saxophone (Figure 1.2). Her tattooed right hand is trained on these precision instruments, as it is trained on the computer keyboard. Both need oxygen, the breath of life giving life to otherwise inert objects. Transformative motion of this sort is indeed a kind of "arrow dynamics", a dedicated vector at the interface of the human and nonhuman, giving shared meaning to our separate paths. At once indigenous and modern, this arrow dynamics includes speeding in the canoe as well as flying through the air. The latter does not come naturally to humans, but our limitations are sometimes forgiven, as Harjo says in her blog entry, "Ahhhh Saxophone": "you take a breath, all the way from the center of the earth and blow. All that heartache is forgiven. All that love we humans carry makes a sweet, deep sound and we fly a little" (2017, n.p.). One of the marvels of our precarious breath is surely to help fuel such flights, launching them at the intersection of the linguistic, physiological, and atmospheric, bringing together the arts and sciences as oxygen can.

Figure 1.2 Joy Harjo and saxophone.

Notes

1 This article has been revised and reprinted from the original publication of "I can't breathe: The Arts and Sciences of Oxygen", in *American Literary History* 34(1) (2022): 113–125, with permission of the editors.

2 Seventy is a number *The New York Times* comes up with. See Mike Baker, Jennifer Valentino-DeVries, Manny Fernandez, and Michael LaForgia 2020, n.p.

References

Anon. 2021. "Sikh Body Provides Oxygen Concentrators to COVID Patients." *Outlook*, October 5. https://www.outlookindia.com/website/story/india-news-sikh-body-provides-oxygen-concentrators-to-covid-patients/382354

Baker, Mike, Jennifer Valentino-DeVries, Manny Fernandez, and Michael LaForgia. 2020. "Three Words. 70 Cases. The Tragic History of 'I Can't Breathe'." *New York Times*, June 29. https://www.nytimes.com/interactive/2020/06/28/us/i-cant-breathe-police-arrest.html

Begay, Jade. 2021. "How Navajo Physicians Are Battling the COVID-19 Pandemic." *Smithsonian*, March. https://www.smithsonianmag.com/science-nature/healers-navajo-nation-pandemic-covid-19-180976994/

Brady, Benjamin R., and Howard M. Bahr. 2014. "The Influenza Epidemic of 1918–1920 Among the Navajos: Marginality, Mortality, and the Implications of Some Neglected Eyewitness Accounts." *American Indian Quarterly* 38 (4): 459–491.

Brown, Alex. 2021. "In Hard-Hit Indian Country, Tribes Rapidly Roll Out Vaccines." *Pew Trusts*, February 9. https://www.pewtrusts.org/en/research-and-analysis/blogs/stateline/2021/02/09/in-hard-hit-indian-country-tribes-rapidly-roll-out-vaccines

Chabba, Seerat. 2021. "COVID: India's Sikh Community Steps Up Amid Pandemic." *Deutsche Welle*, June 2. https://www.dw.com/en/india-covid-sikhs/a-57756219

Cohen, Matt. 2009. *The Networked Wilderness: Communicating in Early New England*. Minnesota: University of Minnesota Press.

Cohen, Matt, and Jeffrey Glover, eds. 2014. *Colonial Mediascapes: Sensory Worlds of Early America*. Nebraska: University of Nebraska Press.

Currizzo, Chelsea. 2020. "COVID Mortality Twice as High among Native Americans than Whites." *US News*, December 10. https://www.usnews.com/news/health-news/articles/2020-12-10/covid-mortality-twice-as-high-among-native-americans-than-whites

Flaherty, Julia. 2019. "For this Navajo Doctor, Listening is Medicine." *Tufts Now*, April 15.

Frayer, Lauren. 2021. "Why is India Running Out of Oxygen?" *NPR*, May 5. https://www.npr.org/sections/goatsandsoda/2021/05/05/989461528/why-is-india-running-out-of-oxygen

Garcia, Edgar. 2020. *Signs of the Americas: A Poetics of Pictography, Hieroglyphs, and Khipu*. Chicago: University of Chicago Press.

Gibbens, Sarah, and Amy McKeever. 2020. "How Breathing Wildfire Smoke Affects the Body." *National Geographic*, September 15. https://www.national

geographic.com/science/article/how-breathing-wildfire-smoke-affects-the-body
Glover, Jeffrey. 2014. *Paper Sovereigns: Anglo-Native Treaties and the Laws of Nations*. Philadelphia: University of Pennsylvania Press.
Harjo, Joy. 2002a. Introduction to *How We Became Human: New and Selected Poems: 1975–2001*, xvii–xix. New York and London: W.W. Norton & Company.
Harjo, Joy. 2002b. "Ah, Ah." In *How We Became Human: New and Selected Poems: 1975–2001*, 185. New York and London: W.W. Norton & Company.
Harjo, Joy. 2008. "For Alva Benson, And for Those Who Have Learned to Speak." In *She Had Some Horses*, 8. New York and London: W.W. Norton & Company.
Harjo, Joy. 2012. "Poet Joy Harjo Talks about Her Tattoo." *Writers Institute Blog*, October 31. http://nyswiblog.blogspot.com/2012/10/poet-joy-harjo-talks-about-her-tattoo.html
Harjo, Joy. 2017. "Ahhhh Saxophone." August 1. https://www.joyharjo.com/blog
Harjo, Joy. 2020. "Introduction." In *When the Light of the World was Subdued: Our Songs Came Through: A Norton Anthology of Native Nations Poetry*, edited by Joy Harjo, 1–10. New York and London: W.W. Norton & Company.
Hedgpeth, Dana, Darryl Fears, and Gregory Scruggs. 2020. "Indian Country, Where Residents Suffer Disproportionately from Disease, Is Bracing for Coronavirus." *Washington Post*, April 4. https://www.washingtonpost.com/climate-environment/2020/04/04/native-american-coronavirus/
Herschlag, Allison. 2020. "The Long Distance Harm Done by Wildfires." *BBC*, August 23. https://www.bbc.com/future/article/20200821-how-wildfire-pollution-may-be-harming-your-health#:~:text=The%20smoke%20has%20even%20been,cities%20as%20far%20away%20Seattle&text=But%20smoke%20from%20wildfires%20can,compared%20with%20average%20air%20pollution
Howley, Elaine K. 2021. "How to Talk to Someone who's Hesitant to Get the COVID-19 Vaccine." *USNews*, May 12. https://www.usnews.com/conditions/coronavirus-and-your-health/articles/how-to-talk-to-someone-whos-hesitant-to-get-the-covid-19-vaccine
Institute of American Indian Arts. IAIA. "About." n.d. https://iaia.edu/about/
Kaur, Harmeet. 2021. "Tribal Health Providers Have Figured Out the Key to COVID-19 Success." *CNN*, February 26. https://www.cnn.com/2021/02/09/us/tribal-health-providers-covid-vaccine-trnd/index.html
Keith, Tamara. 2021. "A New Focus Group Investigated How People Have Overcome Vaccine Hesitancy." *NPR*, May 3. https://www.npr.org/2021/05/03/993205279/a-new-focus-group-investigated-how-people-have-overcome-vaccine-hesitancy
King, James. 2021. "Cherokee Nation Opens Vaccine to Everyone in 14-country Reservation Area." *KTUL*, March 11. https://ktul.com/news/local/cherokee-nation-opens-vaccine-to-everyone-in-14-county-reservation-area
McFarling, Usha Lee. 2020. "How the Cherokee Nation has Curtailed the Pandemic." *PBS Newshour*, November 24. https://www.pbs.org/newshour/health/how-the-cherokee-nation-has-curtailed-the-pandemic
Mudd, Cassidy. 2021. "Cherokee Nation in Phase 3 of Vaccine Distribution." *KTUL*, February 26. https://ktul.com/news/local/cherokee-nation-in-phase-three-of-vaccine-distribution-plan

Pellegrini, Maura, John Pouncett, Mandy Jay, Mike Parker Pearson, and Michael P. Richards. 2016. "Tooth Enamel Oxygen Isoscapes Show a High Degree of Human Mobility in Prehistoric Britain." *Nature*. October 7. https://www.nature.com/articles/srep34986

Porterfield, Carlie. 2021. "Here's Why the Navajo Nation Is Beating Out Every State in Coronavirus Vaccine Rollout." *Forbes*, April 4. https://www.forbes.com/sites/carlieporterfield/2021/04/04/heres-why-the-navajo-nation-is-beating-out-every-states-coronavirus-vaccine-rollout/?sh=109b1e9611cf

Rasmussen, Birgit Brander. 2012. *Queequeg's Coffin: Indigenous Literacy and Early American Literature*. Durham, NC: Duke University Press.

Riebeek, Holli. 2005. "Paleoclimatology: The Oxygen Balance." *NASA Earth Observatory*, May 6. https://earthobservatory.nasa.gov/features/Paleoclimatology_OxygenBalance/oxygen_balance.php#:~:text=The%20exact%20oxygen%20ratios%20can,salinity%20of%20the%20oceans%20worldwide

Siegler, Kurt. 2021. "Why Native Americans Are Getting the COVID-19 Vaccines Faster." *NPR*, February 19. https://www.npr.org/2021/02/19/969046248/why-native-americans-are-getting-the-covid-19-vaccines-faster

Silverman, Holly, Konstantin Toropin, Sara Sidner, and Leslie Perrot. 2020. "Navajo Nation Surpasses New York State in the Highest COVID-19 Infection Rate in the US." *CNN*, May 18. https://www.cnn.com/2020/05/18/us/navajo-nation-infection-rate-trnd/index.html

Singh, Gundeep. 2021. "Meet the Six year old Kid Helping India Breathe, Through Social Media!" *United Sikh,* May 24. https://unitedsikhs.org/meet-the-6-year-old-kid-helping-india-breathe-through-social-media/

Stead Sellers, Frances. 2019. "It's Almost 2020 and 2 Million Americans Still Don't Have Running Water." *Washington Post*, December 11. https://www.washingtonpost.com/national/its-almost-2020-and-2-million-americans-still-dont-have-running-water-new-report-says/2019/12/10/a0720e8a-14b3-11ea-a659-7d69641c6ff7_story.html

Theobald, Brianna. 2019. "A 1970 Law Led to the Mass Sterilization of Native American Women." *Time*, November 28. https://time.com/5737080/native-american-sterilization-history/#:~:text=Over%20the%20six%2Dyear%20period,numbers%20were%20actually%20even%20higher

Treisman, Rachel. 2021. "Outpacing the U.S.: Hard Hit Navajo Nation Has Vaccinated More Than Half of Adults." *NPR*, April 26. https://www.npr.org/sections/coronavirus-live-updates/2021/04/26/990884991/outpacing-the-u-s-hard-hit-navajo-nation-has-vaccinated-more-than-half-of-adults

US Department of Health and Human Services. 2021. *Office of Minority Health*. Diabetes and American Indians/Alaska Natives. https://minorityhealth.hhs.gov/omh/browse.aspx?lvl=4&lvlid=33#:~:text=American%20Indian%2FAlaska%20Native%20adults,whites%20to%20die%20from%20diabetes

Whyte, Kyle Powys. 2017. "Indigenous Climate Change Studies: Indigenizing Futures, Decolonizing the Anthropocene." *English Language Notes* 55: 153–162.

Whyte, Kyle Powys. 2018. "Critical Investigations of Resilience: A Brief Introduction to Indigenous Environmental Studies and Science." *Daedalus* 147: 136–147.

2 "One world or none"

Planetary nuclear precarity and anti-nuclear cosmopolitanism[1]

Pramod K. Nayar

Introduction: anti-nuclear discourses/campaigns

A nuclearized planet is a planet in permanent peril. So how does one speak of a planet in peril? In this chapter, I examine anti-nuclear discourses that construct such a planetary precarity.

In *After Fukushima*, Jean-Luc Nancy (2015) argues, "the spread or proliferation of repercussions from every kind of disaster hereafter will bear the mark of that paradigm represented by nuclear risk" (3). Planetary precarity is planetary *nuclear* precarity, as Nancy indicates. In this chapter, taking this sense of planetary precarity as a given and as a point of departure, I examine the discourses *against* nuclearization. The chapter makes the argument that anti-nuclear campaigns have always been cosmopolitan in origin, scope, and emphasis as a counter to the rise of planetary nuclear precarity.

I introduce some of the earliest documents and cultural texts that highlight the planet in peril. The chapter has four sections. In the first, "One world or none", I begin with Dexter Masters's and Katharine Way's essay collection *One World or None* (1946), to which Albert Einstein, Robert Oppenheimer, Hans Bethe, and other makers of the atomic bomb contributed, and the 1946 film of the same title. In Section Two, "The (under) mining of land", I move on to campaigns against uranium mining in Australia and India, and their documentary film texts. In Section Three, I study texts such as Michael Marder's and Anaïs Tondeur's (2016) photo essay, *The Chernobyl Herbarium*, Rebecca L. Johnson's (2015) book *Chernobyl's Wild Kingdom*, and the work of Cornelia Hesse-Honegger (2008) for the discourse of the nuclearized nonhuman. In the final section, I speak of nuclear cosmograms.

One world or none

In Keiji Nakazawa's ten-volume autobiographical graphic novel of a Hiroshima survivor, *Barefoot Gen* (new English translation 2004–2010),

DOI: 10.4324/9781003383598-4

the eponymous protagonist Gen, eight years after the Hiroshima bombing, admits after reading a news account, that there has been a "jump in cases of leukaemia among A-bomb survivors": "And it terrifies me to think that I could suddenly come down with radiation symptoms too. It makes me feel so helpless" (2009, 10, 165). On the pages following this incident, Nakazawa draws the city's roadsides: buildings and electric posts with posters: "March against the Soviet H-bomb test: Support Pope Pius XII's Proposal to ban Nuclear Weapons" (167). Hiroshima's survivors and victims are protesting against a global threat, in the form of the USSR's atomic bomb programme and, implicitly, the arms race. Spread across time (generations) and space, the nuclear effect is evenly distributed, as Nakazawa implies. That is, this is a sense of precarity derived from a nuclear event at one site and on one day but affecting a much larger region and across time, as Debjani Ganguly (2020) observes in writing about James George's (2006) novel, *Ocean Roads*:

> While the novel's present is 1989 in New Zealand, its historical frame is haunted by anniversaries of nuclear and chemical warfare – the testing of the atomic bomb in the deserts of New Mexico; the bombing of Pearl Harbor; Hiroshima and Nagasaki; the VJ Day celebrations of August, 1945; and the 1969 napalm gas attack in Vietnam. Each of these is captured through the photographic lens of one of its key characters, Etta, a Maori woman and a Pulitzer Prize winner whose filial bonds stretch across these global histories of irradiated violence.
>
> (2020, 437)

Ganguly's phrase, "global histories of irradiated violence", emblematizes the nuclear humanities' principal concern: planetary precarity. In such circumstances, there are no "isolated" nuclear incidents or accidents.

Planetary precarity is engendered by the "nuclearization of the planet", where all life forms and the non-living are subject to risk, and even distant spaces and elements – soil, earth, and water – carry toxins from Chernobyl, Fukushima, and nuclear tests. The toxins from Chernobyl's Reactor 4, we now know, spread as far as the UK, and had reached Sweden in less than 24 hours. Such a precarity acknowledges in an ironic way the "one world or none" stance of the scientists and philosophers – from Einstein to Oppenheimer – in the Masters and Way 1946 volume, *One World or None*, a pioneering text of the anti-nuclear campaign. Planetary precarity produces the advocacy and activist side of an atomic internationalism – or "one worldism" – that originated in the aftermath of World War Two, albeit fuelled by a fear of communism and situated within the context of the Cold War (Andren 2020, 3–4).

At events such as the 1995 "International Symposium: 50th Anniversary of the Atomic Bombs" with Hiroshima's survivors alongside the downwinders of Nevada and the survivors of *Lucky Dragon* (the Japanese fishing boat that was caught in the Bikini Atoll test of the March 1, 1954) fallout, the "*hibakusha* ('survivors of atomic attacks on Hiroshima and Nagasaki') from all over the world" came together, writes Ōishi Matashichi (2011, 77), gesturing at a planetary mobilization of anti-nuclear sentiment.

In his essay "If the Bomb gets out of hand", which features in *One World or None*, Philip Morrison (1946), the US physicist who worked on the Manhattan Project and later visited Hiroshima to study the effects of the bombing, speculates on the possible effects if the atomic bomb was to be dropped on Manhattan. Morrison concludes with:

> If the bomb gets out of hand, if we do not learn to live together so that science will be our help and not our hurt, there is only one sure future. The cities of men on earth will perish.
>
> (6)

In the same volume, arguably the most famous physicist of all time, had this to say: Albert Einstein (1946) opens his essay, "The way out", with:

> The construction of the atom bomb has brought about the effect that all the people living in cities are threatened, everywhere and constantly, with sudden destruction. There is no doubt that this condition has to be abolished if man is to prove himself worthy, at least to some extent, of the self-chosen name of *homo sapiens*.
>
> (76)

Oppenheimer (1946) in the same volume states:

> The vastly increased powers of destruction that atomic weapons give us have brought with them a profound change in the balance between national and international interests. The common interest of all in the prevention of atomic warfare would seem immensely to overshadow any purely national interest, whether of welfare or of security.
>
> (25)

The concluding essay by the Federation of Atomic Scientists states: "*The nations can have atomic energy, and much more. But they cannot have it in a world where war may come*" (78; emphasis in original).

In the documentary film directed by Philip Ragan (1946) *One World or None*, produced by the National Committee on Atomic Information with the assistance of the Federation of American Scientists, as the credits indicate, the opening images show the globe rotating inside the letter O in the words "One", "World", and "Or", which appear one after another, before the word "None" appears. After "None" appears, a small bright speck moves swiftly down from the top of the screen and smashes into the "O" and the globe inside it. Then an explosion rips across the screen and there is only a bright light, with nothing else visible. When the voice-over begins, we are first told of the transnational, and cosmopolitan nature of the scientific discoveries and theories that enabled the creation of the atomic bomb: the German physicist Einstein; the Dane, Neils Bohr; the American, Herbert L. Anderson; the New Zealander, Ernest Rutherford; the Austrian, Lise Meitner; and so on. The commentator says: "no one scientist or nation is responsible". On the screen, the major elements involved in the fission process appear and in the backdrop the various national flags appear as thumbnails. The voice-over speaks of the "pooling" of knowledge. Then, the story shifts to the possibility of an American city such as New York being bombed. A flash of light with the skull inside appears. The skull is placed in the middle of concentric circles that are animated, flashing, and implying the expansion of the range of destruction.[2] Atomic energy, it concludes, can be for the benefit of all, in all nations, provided it is free from the menace of war.

Neils Bohr's (1950) letter to the UN used the definition of the term "cosmopolitan" when he wrote: "the proper appreciation of the duties and responsibilities implied in world citizenship is in our time more necessary than ever before" (n.p.). The responsibility, he said, ought to be directed at an "open" world:

> An open world where each nation can assert itself solely by the extent to which it can contribute to the common culture and is able to help others with experience and resources must be the goal to be put above everything else. Still, example in such respects can be effective only if isolation is abandoned and free discussion of cultural and social developments permitted across all boundaries.
>
> (n.p.)

Written in the wake of the Bikini tests (24 tests between 1946 and 1958), the Einstein-Russell manifesto of 1955 (to which scientists/physicist Max Born, Frederic Joliot-Curie, Linus Pauling, and others also added their signatures) begins by calling the signatories "not as members of this or that nation, continent, or creed, but as human beings, members of the species Man". It calls for a halt to nuclear weapons production but also to war. It urges everyone to think in terms of universals, including universal destruction:

> There lies before us, if we choose, continual progress in happiness, knowledge, and wisdom. Shall we, instead, choose death, because we cannot forget our quarrels? We appeal as human beings to human

> beings: Remember your humanity, and forget the rest. If you can do so, the way lies open to a new Paradise; if you cannot, there lies before you the risk of universal death.
>
> (1955, n.p.)

Thus, in the texts of the first decade (1945–1955) after Hiroshima, the emphasis was on the international control of nuclear power. They do not advocate ending the research into or developing nuclear power – just a mechanism to control its use. And this control, the texts and film suggest, is the task of the world itself. All of this implies an anti-nuclear, global, and "cosmopolitan" responsibility to save the world.

In celebrated documentaries such as Terre Nash's *If You Love This Planet* (1982), the protagonist, Helen Caldicott, an anti-nuke activist, goes into great detail about the dangers of nuclearization. But more importantly, for our purposes, she refuses to localize nuclear harm. Drawing attention to Hiroshima, she says, "in Hiroshima, there was an outside world to come to help. There will be nobody [in the case of an all-out nuclear war]". She then begins an inventory of mass deaths: how many Americans will die, then she adds, also Canadians, Russians, Britons, Europeans, and most of the Chinese. Caldicott signals the likelihood of planetary precarity and global extermination. The pointlessness of assuming a limited war, localized threat, and circumscribed destruction is highlighted in all its absurdity when she declares, "this [nuclear war] is not a war. This is extermination". With even greater rhetorical flourish, Arundhati Roy (2002) writes of the 1998 Pokhran nuclear tests, carried out in Pokhran, Rajasthan, in India:

> If only nuclear war was the kind of war in which countries battle countries, and men battle men. But it isn't. If there is a nuclear war, our foes will not be China or America or even each other. Our foe will be the earth herself [...]. When everything there is to burn has burned and the fires die, smoke will rise and shut out the sun. The earth will be enveloped in darkness. There will be no day – only interminable night [...]. The nuclear bomb is the most anti-democratic, anti-national, anti-human, outright evil thing that man has ever made.
>
> (6–7)

Roy's rhetoric resonates with other anti-nuclear campaigns and their discourses.

The dramatization of the possible destruction of the cities of New York or San Francisco constitutes the "fabulously textual" style (as Jacques Derrida [1984, 23] called it), and structures the prognosis of a nuclear future. Alternating with the archival footage of the Hiroshima destruction, the animation of buildings falling and burning in films like *One World or None* erodes the documentary realism of anti-nuclear activism. The fabulously textual mode has recourse to a figural realism – whether this is in the animated projection/fantasy of a future war in the films or in the metaphor of "way [...] to a new Paradise" in the Einstein-Russell manifesto. In David Bradbury's 2007 film *A Hard Rain*, the effects of radon gas – one of the materials that emerges in uranium mining – are presented in provocative ways. We are first told how it affects minds, families, and communities – and for the latter, the film shows Aboriginal families walking through the open land – before showing us a sleeping infant. As the voice-over (from Dr. Rosalie Bertell, an epidemiologist) tells us, the invisible, odourless gas gets into the body, and the film shows ghostly, wispy cloud-like vapours floating around the baby. Then, it shows the animation of lungs inhaling the gas, causing "teratogenic damage". In this analysis, the film moves inward – as opposed to the catastrophic realism that moved away from the hypocentre of the bomb farther and farther out. First the landscape, then communities and families, then the sleeping infant and finally the lungs. Also, the register is shifted from the aesthetics of the prospect view – the uranium mining area, the landscape – to the intimately textual detail of the infant accompanied by the ghostly, and finally, the uncanny images in which the lungs appear as actively animated. Bradbury, like Niels Bohr in his letter to the UN and the contributors and editors of *One World or None*, seeks to emphasize the future-catastrophic by moving away from the realist mode of the documentary.

The (under)mining of land

In the 2009 documentary film, *Uranium: Is It a Country?*, directed by Kerstin Schnatz et al., Reg Dodd begins by telling us that in an area of about 150 kilometres in Australia, there are three to four "big mines". He informs us of the Aboriginals – he belongs to the Arabana peoples – who do not go near such mining places, because they believe they are "poisonous". In the same film, Dave Sweeney of the Australian Conservation Foundation suggests that given the role uranium plays in war, the best thing for the world to do is to "keep the genie in the bottle". From Dodd's point of view, the discourse is of the toxification of the land and from Sweeney's, it is what the land is made to exude or spew out that toxifies the *world* at large. This set of shots is interspersed with (i) the head of the uranium mining corporation who speaks in terms of the profits generated annually from Australia's export of uranium (Australia holds 30–40 percent of the world's uranium deposits) and (ii) an anti-uranium protest march.

Multiple discourses intersect here, from capitalism and its ruthless pursuit of profit to the local communities' ignorance, beliefs, and fear about mining, the role of place/s, and finally, the "saving-the-world" position.

Three domains figure prominently in anti-nuclear discourses: uranium mining, nuclear testing, and nuclear waste disposal: in all these domains, the place is the focus.[3]

In David Bradbury's (1997) *Jabiluka*, emphasis is laid on the fact that the land is the "traditional home of the Mirrar people, and the world's oldest living culture". Speaking about the land, Yvonne Margarula says in the film, "we own the country, our country... black country, not white country". Into this land, which has "sacred sites of significance [to the Aboriginals]", comes the white man who, says the voice-over, "understands the value of uranium in an energy-hungry world". The mining, as the protesters and campaigners point out in films like *Jabiluka* and Ned Lander's (2011) *Dirt Cheap 30 Years On: The Story of Uranium Mining in Kakadu*, disturbed the traditional environment of the Mirrar peoples. In Shri Prakash's (1999) *Buddha Weeps in Jadugoda*, on uranium mining in the Jadugoda region of Jharkhand state of India, the documentary opens with the origin story of the people of the place: the Santhals, Mundas and other Adivasis. The film focuses on their sacred groves and their rituals. A journalist from the community says in the film that the land's resources are "not for our [tribals'] prosperity [...] they have become our curse".

In anti-nuclear documentaries from countries as diverse as Australia and India, then, uranium mining is pitched as the binary opposite of, indeed as antagonistic to, cultural heritage and the values of the local populace and residents. The Aboriginals and *Adivasis* represent one specific set of values. The modern man (white or brown) represents the exact opposite. There is also a discourse of Nature versus culture, where the Aboriginal way of life is, not unproblematically, but perhaps as an instance of a strategic essentialism, valorized for being closer to Nature and less exploitative of it, while modern white man's culture is brutal and exploitative in the extreme. Kakadu, Jabiluka, and Jadugoda represent, in this rhetoric, places older than time, and the very antithesis of the contemporary.

In *Buddha Weeps in Jadugoda*, we are shown scenes of rituals and ceremonies being held in the community. There are also shots of community prayers, feasts, and cultural festivals. In these images, as with scenes from other documentaries of indigenous populations in their everyday lives, the place functions as material rhetoric (Endres and Senda-Cook 2011, 262). The emphasis on local cultural practices, memories, and values of sacralized places in anti-mining campaigns renders the place both material and symbolic, and constitutive of the very rhetoric of protest. The protection of sacred places, as seen in the rhetoric of all these protests, is offered as a counter to nuclear colonialism (and its rhetorical moves, including the rhetoric of nuclear stewardship, in the post-Cold War period) that has been disrespectful of place.[4]

There is another aspect of this material rhetoric of place. The battle is against the local sources of materials that fuel massively catastrophic wars that are often fought elsewhere. The protests, therefore, invoke places from elsewhere as constitutive of protests *here*. As the protestors' placards declare in *Uranium: Is It a Country?*: "stop uranium mining to stop wars". That is, it is not war per se that is being protested, but the backdrop to the manufacture of more and more dangerous weapons, through uranium production and the transnational impact of, say, Australian uranium production on wars in other parts of the world. This is a transnationalization of the material rhetoric of those places that are marked by the presence and extraction of uranium. The other parts of the world are material embodiments of Australian uranium, so to speak. The uranium in an Australian mine is transnationalized, just as a war with nuclear weapons is transnational and planetary in scope. This is the material rhetoric of uranium mining and production.

Nuclearization dismantles places that have for centuries supported indigenous populations. Hence, in most cases within these discourses, there is an overwhelming emphasis on the effects of "nuclear colonialism" (Endres 2009) on indigenous peoples.[5] It is important to note that this modality of colonial domination of the natives in places where natural resources that European capitalism and modernity desired is an extension of older colonial practices as well (see Hecht 2003). Even nuclear alarmism in the rhetoric of Euro-Americana carries vestiges of colonial discourse: in this case, it is constantly reiterated that nuclear weapons are better off in the hands of Europe and America rather than in those of Third World nations (see Gusterson 1999; Biswas 2014). Thus, uranium mining and its deleterious effects born of white erasure of the indigenous has a long history – and this has been the subject of anti-nuclear campaigns since the post-Cold War period.

In *Uranium: Is It a Country?*, Michael Angwin of the Australian Uranium Association speaks of the dollars earned through uranium mining and export (Dodds 2008). The emphasis on profits and national interest on which this discourse operates implicitly calls for sacrifices on the part of those communities affected by the mining operations (Endres 2009, 45). The CEO, Chamber of Commerce and Industry, South Australia, Nigel McBride, in Timothy Large's (n.d.) documentary *Nuclear Wasteland*, admits that there are about 100 nuclear waste sites across the country. Then comes the horrific statement, "probably not stored particularly well [...] really not an ideal position".

In sharp contrast to the CEOs in these films, the community leaders speak of the loss of their lands, their rights, and their memories of the place. Thus, in the case of nuclearization, the creation of mining areas, quarries, and factories renders the place of indigenous occupation and residence into non-places, in the sense that Marc Augé (1995) uses the term. In the process, the indigenous people become non-persons on the sites: they are replaced by engineers and technicians. In effect, the non-place is constructed through the operationalizing of an older discourse: that of *terra nullius*.[6]

In every protest campaign, indigenous peoples speak of their displacement from the land of their ancestors. The white man and his machines have settled into their lands, and desacralized them for ever through two modes. First, the intrusion of men and materials that arrive as material embodiments of a distinct culture and process – extraction of natural resources. Second, the resources from under the ground are actually sold in international markets and to foreign nations. That is, the sacred land with its special significance has been treated as not just *terra nullius* – devoid of people, or nobody's land – but as being on a par with any land with resources that the white man values. Further, by transporting materials from under the soil to different parts of the world, nuclear policy has redistributed the sacred land and fragmented it. The objections of the indigenous people who do not wish to be incorporated into the "new" landscape of mining and extraction are overruled – exactly as settler colonialism operated a century and a half earlier in Australia or Canada.

This is the nuclear *terra nullius*: the operations of the state and the corporation(s) empty the land of its rightful owners and residents but, more importantly, render the region and the land unfit for the future as well. That is, the nuclear *terra nullius* is simultaneously about the past and the future. The land, the dwelling place of the indigenous people for centuries, has been deemed empty of people and full of resources. In other words, nuclear colonialism proceeds with a double movement: of emptying the land of its residents and treating it as "full". The uranium beneath the soil makes the land rich, and the people living on this land of no account. The nuclear *terra nullius*, therefore, makes the people exterior and extraneous to the *matter* beneath their feet. The fact that this same land carries the memories and bones of the residents' ancestors is, likewise, dismissed with the focus exclusively on the minerals. The people are excess, surplus, but the minerals are not. The de facto logic of *terra nullius* is operationalized and made visible in a different mode in the case of nuclear testing and waste disposal sites – and this is documented as part of the art of anti-nuclear campaigns.

Peter Goin (1991) captures the nuclear *terra nullius* in the photographs in the volume *Nuclear Landscapes*. He sets up the sequence of photographs through a photographic textual framing, as Edmund Lee and Shirley Ho (2017) term it. He begins with photographs of the Trinity explosion, the construction of the Hanford Nuclear Reservations, the Marshall Islands,

Nevada tests, and the landscape at the test sites. Goin's introductory section ends with:

> The Trinity Site is now a National Historic Landmark [...]. Bikini and Enewetak atolls are supposedly cleaned-up, and only the bunkers remain. More than 30 years have passed since the last nuclear detonation at Bikini and Enewetak, and the islands are becoming overgrown with vegetation. Many of the Hanford reactors have been decommissioned. Yet the sites are radioactive.
>
> (Goin 1991, 27)

Thus, Goin has already indicated the nature of the landscapes we will see: they are radioactive. Yucca Flat now shows signs of life with Joshua trees on the desert floor in one photograph (1991, 39).[7] Remnants of buildings set up for various purposes in these areas dot the land, and an apparatus such as an army tank in Yucca Flat "remains highly radioactive" (40–41). Elsewhere, in the South Silent Canyon of the Nevada Test Site, the canyon's neighbouring cliffs show signs of accelerated erosion (52–53), the fault lines have shifted on the valley floor of Yucca Flat (54–55), subsidence craters appear in other cases (50–51, 58–59, 62–63). In these pictures, Goin focuses on the geological damage from nuclear tests. (The Yucca Mountain was authorized as America's first high-level nuclear waste repository in 2002, before federal funding was stopped in 2011).

In "Orchard Site", Goin examines the Hanford site. These were originally "successful farming regions", and hence hardly *terra nullius*. The "orchard was cut down after the area was condemned for use in the Hanford Nuclear Reservation", he informs us (Goin 1991, 66). The photograph shows a landscape of scrubs with hundreds of tree stumps (67). It is exactly what it looks like: a site of deforestation, except that the deforestation is nuclear-driven. Goin also photographs cement debris – which is radioactive – at Hanford (70–71). The waste storage pond has water, which is "contaminated with radioactivity" (74–75). There are landscapes with "yellow posts [that] identify buried radioactive waste and potential surface contamination" (76–77).

Three photographs, all titled "Burial Ground", from the Gable Mountain Pond waste burial site, capture a landscape where there are "massive storage tanks containing hundreds of millions of gallons of liquid radioactive waste", Goin writes, "some reports indicate that the tanks are already leaking" (1991, 82; see photographs: 83, 85, 87). The Retention Basin area has "surface contamination" (90–91). The fact that Goin uses the term "Burial Ground" is evocative: many of the lands across the world in which nuclear wastes are interred are also sacred lands for the indigenous peoples of the region.

After covering the USA, Goin turns to the Bikini and Enewetak Atolls. He photographs the massive "crypt" storing "111,000 cubic yards of radioactively contaminated soil" (104–105). In "Nuclear Bunker Complex", we see a patch of green adjacent to the concrete structure. Goin dispels any hope one may take from the lush green when he writes "this area is still radioactive" (1991, 106–107). Debris scattered across the islands is "still radioactive" (116–117). "Radioactive vegetation" covers the bunkers elsewhere on the atoll (130–131). In his commentary on "Bunker" (136–137), Goin summarizes the nuclear *terra nullius*:

> According to a Department of Energy survey report (April 1989), if islanders were to resume living here, they would be receiving a 20-millirem dose of radiation over and above normal background radiation. The surface soils of Enjebi Island are contaminated with residual levels of plutonium (half-life of 240, 360 years) and americium, which could be re-suspended by the wind and present an inhalation hazard.
> (136)

In short, the island, if peopled, poses a threat that is likely to last a very long time as the half-life span indicates. For his last photograph, "Tide Pool", Goin writes: "The Marshall Island sites of Bikini and Enewetak are still radioactive" (144).

Goin captures empty lands. However, as his comments indicate – and it is common knowledge too, now – the Bikini and Enewetak were inhabited spaces, as were lands in the Hanford area. These had been *emptied* and the natives relocated to facilitate the tests. This is the first iteration of the nuclear *terra nullius*, and it is deeply disturbing. The second iteration is even more frightening. Goin's landscapes have been contaminated for a very long time. The emphasis on leakage of radioactive materials, the strewn radioactive debris, and finally, the very nature of radioactivity which is unlikely to cease for the term of humanity's natural life – with half-lives of 240, 360 years – ensures that the area has to remain *terra nullius*, unless humans wish to absorb more than the acceptable limit of radiation if they return to the atolls. In other words, it is a permanent *terra nullius*.

The nuclearized nonhuman

Goin photographs a landscape of coconuts on Eneu Island, Bikini Atoll. However, these are unsafe. Goin says, "the residual levels of cesium 137 (half-life of 30 years) absorbed into the coconuts [...] are still too high for human consumption to be safe" (1991, 124–125). He titles the image, "Coconut Graveyard". The last photograph, "Tide Pool", shows electric cables intertwined with coral in the Bikini and Enewetak Atolls. Goin tells us: "even the best cleanup efforts could not remove all the electrical wire, pipe, and assorted smaller objects. Some were welded naturally into the coral in tidal areas throughout Bikini and Enewetak atolls, creating an unnatural beauty" (144). What remains of the "natural" in the collection of coconuts is the key question.

In an early essay, "Effects of Nuclear Testing on Desert Vegetation", Lora Shields and Philip Wells (1962), observing the vegetation in Yucca and French-man flats in the northern Mohave Desert, comment:

> It might be expected that after 8 years of nuclear testing very little vegetation would remain. On the contrary, except for the usual barren playa, no part of this basin lacks flowering plants. At a distance of 2 mi from most "ground zeros" the vegetation shows no visible effects of weapon testing. Grotesque Joshua trees [...] relieve the gray monotony of the dominant shrubs, hopsage [...] and blackbrush.
>
> (38)

They note that "'pioneer' species – non-native to the region – appear to be invading at the perimeters of the totally denuded portions of certain ground-zero areas" (40). This in itself does not strike observers as odd: that the native vegetation has entirely disappeared and that the so-called return of nature is an effect of human engineering, even when species change was not the intended effect of this engineering. In other words, the replacement of species was a biocultural phenomenon induced by the nuclear tests in the region.[8]

In the same vein, Paul Boyer (1998) cites an example of the media discourse around the nonhuman effects of nuclear testing:

> In its August 11, 1951, issue, to cite only one of hundreds of examples, *Collier's* magazine published "Patty, the Atomic Pig." The article was based on an actual incident in which a piglet that was part of the "Noah's Ark" of goats, pigs, rats, and other experimental animals assembled for the July 1946 Operation Crossroads nuclear test at Bikini atoll was later found swimming in the radioactive waters of Bikini lagoon. The *Collier's* story [...] [i]llustrated with cute drawings, [...] imagined Patty's thoughts before the blast [...] and her adventures afterwards [...] Patty not only survives (no scary radiation-exposure hazards here!) but grows to be a six-hundred-pound porker under the benevolent care of kindly scientists at the Naval Medical Research Institute at Bethesda, Maryland, and ends her days as a coddled exhibit in a zoo. Sugar-coated propaganda like this, part of a mountain of material in the media that reinforced the government's version both of the 1945 bombings and of Washington's subsequent nuclear program, served to deflect

> and neutralize serious scrutiny of the meaning and the implications of atomic weaponry, past, present, or future.
>
> (253–234)

Such reportage was meant to reassure the public about the nuclear power being unleashed. In sharp contrast, for the locals, the visible signs of radiation sickness were present on the plants and animals of their region. In the Cedar City area, for instance, one sheep farmer records:

> We started noticing white spots coming on [black sheep's] wool, on the top of their backs. We had a black horse there at camp, and pretty soon he had white spots come on his back and on his rump and all over, from this fallout, I guess.
>
> (cited in Fox 2014, 61)

Sarah Fox (2014) documents several such case histories of the consequences of nuclear testing on animals in the American west. Peter Goin photographed an enclosure once used to pen animals in order to study the impact of the nuclear explosions on them (1991, 47).

Photographs of pine trees (145 of them), specially planted in the Nevada test site to monitor the impact of nuclear explosions, show them swaying hard under the impact. Elsewhere, in regions like Puerto Rico, experimentation with radiation was conducted by the USA, so that these regions became "nuclear laboratories" (DeLoughrey 2012, 172). These places became, as Elizabeth DeLoughrey (2012) has argued, both laboratory and ecosystem, where the latter was deemed to be a "closed" one. She writes: "Rethinking the ways in which science used the isolated island concept to produce some of the most apocalyptic technologies on earth challenges both the assumption of the primitive ahistorical island and what constitutes the laboratory itself" (175). Indeed, echoing DeLoughrey's argument is Grigoriy Medvedev (1987) in *Chernobyl Notebook*, who speaks of the ancient land of the area and the "sense of the primordial" (26). Into this primordial area would arrive the laboratory, the manufacturing unit and radiation.

In the series *Chernobyl* (dir. Johan Renck 2019), much of Episode Four is concentrated on the killing of animals – cats, dogs, milch cattle – in the area, because they had imbibed radiation and were potentially lethal. Rebecca Johnson (2015) in *Chernobyl's Wild Kingdom* documents the effects of the disaster on animals – years after the explosion. She maps a post-natural scene. There was, contrary to expectations, a return of life forms in the Exclusion Zone:

> Beyond the fence, the vegetation is lush and green. In some places, trees grow right up to the edge of the road. Their branches arch overhead

> to form a leafy canopy so dense it filters out some of the sunlight and casts deep shadows all around. Clumps of grass and weeds sprout from cracks and potholes in the crumbling asphalt, and wildflowers bloom along the roadside.
>
> (18)

But it is not the lifeforms per se that Johnson is interested in: it is the nature of their nature, so to speak, their habitats and the food they consume. Commenting on the animal photographs by Sergey Gaschak, deputy director Science, Chernobyl Center for Nuclear Safety, she writes:

> more than fifteen hundred species of plants, mosses, and lichens are now growing in what amounts to ground zero of the nuclear accident [...]. Gaschak has spent a lot of time looking at wildlife [...] has seen moose with newborn calves crossing marshes, knee-deep in mud saturated with the remains of nuclear fallout. He has seen wild boars rooting for mushrooms around the gnarled roots of trees that have been growing in radioactive soil for years. He has even spotted birds nesting in the nooks and crannies of the concrete sarcophagus that surrounds Reactor Number 4. (22–23).

Many of Gaschak's photos could easily pass for snapshots of a pristine wilderness. How could a place so contaminated by nuclear fallout become a seeming paradise for wildlife?

Johnson's interest lies in the flourishing of life in the radioactive zone, terming the entire region a "radioactive nature preserve" (25). After admitting that the absence of the human element enabled the animals to return and flourish, she writes:

> the Exclusion Zone remains the most radioactive environment on Earth. It was so contaminated by nuclear fallout that it won't be completely safe for people to live there for hundreds of years. Hot spots such as the Red Forest will likely be uninhabitable by humans for an even longer period of time. Yet, the Zone's animals don't seem to be aware of the danger all around them.
>
> (25–26)

She then sets out to explore the effects of consuming radioactive plants and life forms on these animals. Initial tests demonstrated no visible effects, leading to the conclusion that the radiation did not harm the animals. Later tests on barn swallows showed they were all radioactive, and many showed demonstrable effects like albinism (38). Shortened life spans were also recorded (40). In some species, they found mutations (45).

In a later chapter, Johnson moves out of Chernobyl, into Fukushima. Scientists who had observed Chernobyl went to Fukushima and found the same results:

> Timothy Mousseau and Anders Møller have visited Fukushima's Exclusion Zone every summer to carry out many of the same types of studies they have been doing in Chernobyl. They have seen similar signs of radiation damage in animals from both sites. In places with high levels of radiation, for example, the researchers found significantly fewer than normal numbers of birds, spiders, and insects. And they have spotted barn swallows with patches of white feathers.
>
> (51)

Johnson concludes:

> In short, life has carried on. While scientists continue piecing together the full story of what is happening in the shadow of Reactor Number 4, the animals that creep and scamper and swim and soar through the abandoned lands around it will go on living as best they can in the radioactive kingdom they call home.
>
> (55)

Johnson's text is not in the same category as avowedly anti-nuclear campaign and propaganda texts but can be clubbed with them for what it sets out to do: use images and text to alert us to the nonhuman consequences of nuclear power, a power that is invisible and yet all pervasive. The animals "flourishing" in Chernobyl are life forms that have imbibed and continue to imbibe radiation. Rewilding and refaunation (Grebowicz 2021, 59–60) are "radiant" rewilding.

Cornelia Hesse-Honegger, an illustrator at the University of Zurich, collected specimens of insects in areas with nuclear power plants and in regions in direct line of the toxic cloud from Chernobyl: Sweden and Switzerland, and from the Chernobyl region. Over time, Hesse-Honegger collected and drew over 16,000 true bugs from 25 nuclear sites around the world. Later, she took live samples from areas impacted by the Chernobyl cloud, breeding *drosophila* (fruit flies) in her home to observe abnormalities in the offspring. With Peter Wallimann, she published a 40-page study "Malformation of True Bug (Heteroptera): a Phenotype Field Study on the Possible Influence of Artificial Low-Level Radioactivity", with images of these deformed insects as a result of radiation, in the journal *Chemistry & Biodiversity* in 2008. Her other work appeared on Atomic Photographers Guild website (https://atomicphotographers.com/cornelia-hesse-honegger/). The artistic renderings attracted both scientific and public annoyance and fascination, as a Smithsonian essay (Thompson 2014, n.p.) noted.

Hesse-Honegger focuses on the loss of bodily similitude to the normative form of the insect. The deformity implies the presence of something not-quite-life: the toxin. In the process, she captured the transformation of Nature: insects breeding after ingestion of toxins, their genetic material modified and, more seriously, passed on too. As Hesse-Honegger states early on: "Sensing that Nature was more and more endangered, I gradually developed the notion that mutated laboratory flies were physically rendered prototypes of our destructive behavior, materializing the future of Nature" (Hesse-Honegger and Wallimann 2008, 500). She concludes:

> [M]y field studies show that a significant percentage of European true bugs, living in their specific habitats, are highly disturbed, not only in terms of the actual number of individuals affected, but also regarding the quality and severeness of malformation [...] it confronts us with ethical questions regarding Nature and Life in general. From the scientific point of view, it is necessary 1) to investigate the long-term effects of low-level artificial radiation; 2) to look at the radionuclide-specific effects on plants and animals; and 3) to reconsider the current threshold values for radioactive immission.
>
> (Hesse-Honegger and Wallimann 2008, 538)

Hesse-Honegger is speaking of a disruption in the order of Nature, as evidenced in the mutated bugs. Her insistence on the ethical questions implies that nuclear energy has interfered with the order(s) in Nature – since the orders in Nature are themselves polyphonous, as Lorraine Daston has argued (2018) – and hence, as a biocultural phenomenon, needs to be investigated and perhaps stopped.

In his account of the "Lucky Dragon" incident (the US nuclear test, Bravo, at Bikini, March 1, 1954), Ōishi Matashichi (2011) reports how contaminated tuna fish from the incident were called "weeping fish" because each tested specimen had counts of 2,000–6,000 on the Geiger counter, indicating massive radiation exposure (29). Addressing a different form of life, Michael Marder (2016) writes in *The Chernobyl Herbarium*:

> Chernobyl's human survivors are the scraps of radiation's after-life, which severely limits life expectancy as a consequence of external and, in many cases, ongoing internal exposures. Plants grown in contaminated soil are, likewise, a finite afterlife of radiation. Strontium-90 accumulates in vascular vegetal tissues, whereas cesium-137 is distributed throughout a plant, due to its similarity to potassium.
>
> (38)

The photograms composed by Anaïs Tondeur are, in Marder's terms, "an emblem – being thrown into, *emballein* – of afterlife" (2016, 38).[9]

They serve as testimonies and material witnesses to the invisible contamination of the land.

Plant life grows in the crevices of concrete structures that are radioactive (Goin 1991, Johnson 2015). Animals consume plants and other creatures and imbibe radioactive nuclides. The plants are captured *with* their "inner" glow, so to speak. The rhetoric of such art work focuses not on the visible depredations and denudations of the nuclearized land, but on the lushness of life. In a shift away from the traditional frugal economies of environmentalism, these texts show life thriving in the midst of radiation. There is visible pleasure, Johnson implies, in seeing the uninhibited growth of plants and animals. Life then has performed its liveliness in the midst of inhospitable environments. The very notion of an "Exclusion Zone" is absurd given that nothing is excluded from the radiation that *leaves* the Zone. The art work draws attention to precisely this condition of boundarilessness in the action of radiation. Radiation is the agential act of the non-living upon the living. This flourishing of life in the Exclusion Zones is itself a troubling index of a future not yet here, but palpable. These texts are countable as anti-nuclear texts precisely because they all signal the horror of a nuclearized planet in which there is no possibility of an Exclusion Zone.

The nuclear cosmograms

The symbol of radiation hazard is now a common visual vocabulary for the planet. Used across the nuclear sites in the world, the symbol is indeed a symbol of planetary precarity. It is a cosmogram. A cosmogram is "a text that results in a concrete practice and set of objects, which weave together a complete inventory or map of the world" (John Tresch, cited in Gabrielle Hecht 2007, 102). In Gabrielle Hecht's (2007) gloss, "Cosmograms concretize: they offer a set of practices, of rituals to enact participation in the world" (107). The symbol for radiation is, I suggest, a cosmogram that enables an understanding of the world, our participation in it, and a set of practices – all of which have to do with the presence of a nuclear threat.

The ritual posting of the symbol at multiple spots around the world brings the world together into the orbit of nuclearization and its attendant risks, potential, and threat. It cautions people against specific areas (see Figure 2.1).[10] The three propeller blades are meant to illustrate the radiation activity of the atom at the centre, but it could also signify the earth in the middle hemmed in by activity of some kind, from all three sides. That is, it need not be the centrifugal (radiating outward from the centre) but a centripetal movement of the danger from the outside enveloping or closing in. As a cosmogram, the radiation hazard symbol is literally, then, a map of the planet, perhaps photographed or drawn from an aerial view and, therefore, the nuclear equivalent of the famous Blue Marble photograph

Figure 2.1 Radiation Hazard Symbol.

of NASA's Apollo 17. The view from above in photography, write Denis Cosgrove and William Fox (2010):

> allows us to plan ahead, to place ourselves in the larger context of the world and map out a course in both space and time. The fundamental utility of the aerial view is why we are so attracted to the examples in the museum – we can't help looking at a view that we know carries such important information.
>
> (11–12)

They elaborate: "aerial photography does remarkably well to reveal and even create patterns at varying scales on the earth's surface. Unless viewed stereoscopically or taken from an oblique angle in sharp shadow, it also flattens the image, emphasizing surface" (100). The earth in the middle appears shrunk ("not to scale"!) and overwhelmed by the blades surrounding it from three sides. It informs us that the earth is *surrounded*.

The second cosmogram (Figure 2.2) is the Doomsday Clock. Designed by Martyl Langsdorf, it was unveiled in 1947 by the *Bulletin of the Atomic Scientists* (a group founded by Einstein and others involved in the Manhattan Project), whose first issue appeared four months after Hiroshima-Nagasaki. The Clock has been taken to indicate how close we are to disaster/midnight. Others claim that the Clock is meant to indicate the current level of risk facing humanity. S. J. Beard (2022), who works in the field of existential risk analysis, writes:

> Its [the Clock's) goal is not to tell us how big the risk facing humanity is, but how well we are doing at responding to that risk. For instance, in 1962 the Cuban Missile crisis is generally agreed to have been the closest the world ever came to nuclear war, but its occurrence did not move the

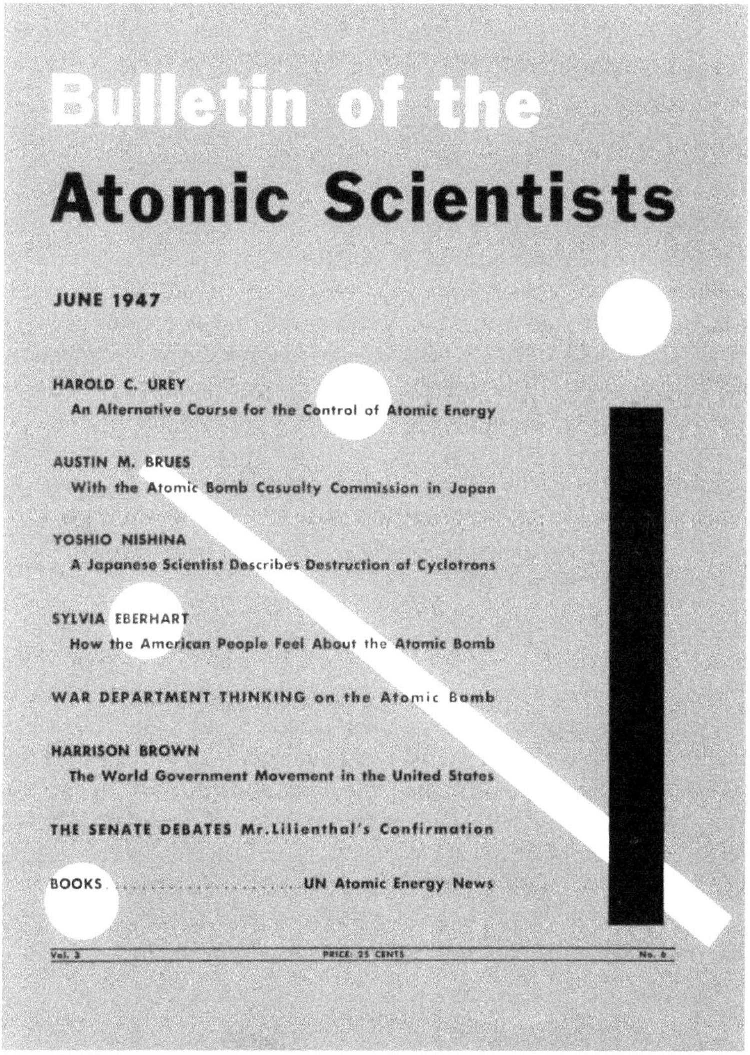

Figure 2.2 Doomsday Clock on the cover of the *Bulletin of Atomic Scientists*, 1947. The clock was designed by Martyl Langsdorf.

> Clock at all. On the other hand, the 1963 Partial Nuclear Test Ban Treaty saw the Clock's hands shift back from midnight an entire five minutes.
>
> (n.p.)

(The Clock, originally designed to reflect the threat from nuclearization, has included climate change in its calculations since 2007.) The Clock is a symbol, not a scientific or quantitative assessment of the risk the planet is embedded in. Rachel Bronson, the *Bulletin* president and CEO explains:

> When the clock is at midnight, that means there's "been some sort of nuclear exchange or catastrophic climate change that's wiped out humanity," she said. "So we never really want to get there and we won't know it when we do."
>
> (quoted in Marples and Ramirez 2022, n.p.)

The Clock signals how close we are to annihilation and the efforts we as a species are making to delay the catastrophic event. Setting the clock back or forward, then, constitutes an awareness and an action plan vis-à-vis planetary annihilation. If the regular clock *shows* the passage of time in the form of the moving hands, the Doomsday Clock shows the time *left* for us as a planet. Then, the Doomsday Clock as a cosmogram is not, unlike a traditional watch or clock, pointing to the local time, or the time at a specific place. Rather, it symbolizes something as unquantifiable and even unimaginable as planetary time: the time of the earth itself and all the matter on it. Taken together, the Doomsday Clock and the radiation symbol signal a planetarity: that wherever you are, your future (time) remains linked to the rest of the planet.

Conclusion: a planet in peril

Discourses of anti-nuclearization have taken multiple forms, genres, and aesthetic modes, as this chapter has hopefully captured. Despite different emphases ranging from the effect on places to the nonhuman, in media forms as diverse as the graphic novel and the animation film and the polemical essay, anti-nuclear discourses remain committed to one overarching theme.

With nuclear power's omniscience and endless (atemporal) effects, we can no longer speaker of *one* site of nuclear tests or bombing, or a test or bombing as an event that occurred at a particular time. There is no such thing as a limited nuclear war. A planet in peril for thousands and thousands of years is a truth the anti-nuclear discourses underscore throughout. To nuke or not to nuke can only be decided if we choose "one world or none".

Notes

1 This essay derives from my study, *Nuclear Cultures: Irradiated Subjects, Aesthetics and Planetary Precarity* (Nayar 2023).

2 The film also traces a history of incremental destruction due to newer and newer weapons in war: the spear killed one man, the cannons killed 88, and so on, in order to argue that the scale of destruction has always been rising.
3 Hecht (2010) traces the genealogy of uranium being transformed from mineral to a "nuclear thing" or "nuclear material".
4 For a study of the nuclear stockpile and stewardship rhetoric, see Taylor and Hendry (2008).
5 Even the *hibakushas*, argue some commentators and novelists, were erased from memories and the land. In *Ocean Roads*, Akiko tells Caleb: "At one end of the building, there was a ward just for them. Even those who had survived had had their names wiped away, by my people, by the Americans. The great, collective forgetting. United only by their bandages" (George 2006, 328).
6 That the concept of *terra nullius* adversely impacts indigenous people is cited by the UN itself. See United Nations, "Doctrines of Dispossession". Also see Fitzmaurice (2007).
7 In 1989, the Department of Energy abandoned its effort to assess the technical suitability of the Yucca Mountain as a potential site for the nation's only civilian high-level nuclear waste repository.
8 Similar instances of life forms thriving in regions of nuclear testing such as Bikini Atoll have often appeared in the mass media. See, for example, Keating (2017) and Sperling (2020).
9 The photograms were "created through the direct imprints of radioactive herbarium specimens, grown in the soil of 'the exclusion zone' by Martin Hajduch of the Institute of Plant Genetics and Biotechnology at the Slovak Academy of Sciences and arranged on photosensitive paper" (Marder and Tondeur 2016, 11).
10 On the radiation hazard sign's origins, see Lloyd Stephens and Rosemary Barrett (1978).

References

Andren, Mats. 2020. "Atomic War or World Peace Order? Karl Jaspers, Denis de Rougemont, Bertrand Russell." *Global Intellectual History* 7 (4): 1–17.

Augé, Marc. 1995. *Non-places: Introduction to an Anthropology of Supermodernity*. New York: Verso.

Beard, S. J. 2022. 'How to Read the Doomsday Clock'. BBC. January 20. www.bbc.com/future/article/20220119-how-to-read-the-doomsday-clock.

Biswas, Shampa. 2014. *Nuclear Desire: Power and the Postcolonial Nuclear Order.* Minneapolis, MN and London: University of Minnesota Press.

Bohr, Neils. 1950. "Open Letter to the United Nations." Accessed March 12, 2022. https://www.atomicheritage.org/key-documents/bohr-letter-un

Boyer, Paul. 1998. *Fallout: A Historian Reflects on America's Half-Century Encounter with Nuclear Weapons*. Columbus: Ohio State University Press.

Bradbury, David, dir. 1997. *Jabiluka*. UK and Australia: Journeyman Pictures Ltd and Frontline Films.

Bradbury, David, dir. 2007. *A Hard Rain*. NSW, Australia: Frontline Films.

Cosgrove, Denis, and William L. Fox. 2010. *Photography and Flight*. London: Reaktion.

Daston, Lorraine. *Against Nature*. Cambridge, MA: MIT Press, 2018.

DeLoughrey, Elizabeth. 2012. "The Myth of Isolates: Ecosystem Ecologies in the Nuclear Pacific." *Cultural Geographies* 20 (1): 167–184.

Derrida, Jacques. 1984. "No Apocalypse, Not Now: (Full Speed Ahead, Seven Missiles, Seven Missives)." Translated by Catherine Porter and Philip Lewis. *Diacritics* 14 (2): 20–31.
Dodds, Klaus. 2008. "Hollywood and the Popular Geopolitics of the War on Terror." *Third World Quarterly*. 29 (8): 1621–1637.
Einstein, Albert. 1946. "The Way Out." In *One World or None*, edited by Dexter Masters and Katharine Way, 76–77. New York: McGraw-Hill.
Endres, Danielle. 2009. "The Rhetoric of Nuclear Colonialism: Rhetorical Exclusion of American Indian Arguments in the Yucca Mountain Nuclear Waste Siting Decision." *Communication and Critical/Cultural Studies* 6 (1): 39–60.
Endres, Danielle, and Samantha Senda-Cook. 2011. "Location Matters: The Rhetoric of Place in Protest." *Quarterly Journal of Speech* 97 (3). 257–282.
Fitzmaurice, Andrew. 2007. "The Genealogy of Terra Nullius." *Australian Historical Studies* 38 (129): 1–15.
Fox, Sarah Alisabeth. 2014. *Down Wind: A People's History of the Nuclear West.* Lincoln: University of Nebraska Press.
Ganguly, Debjani. 2020. "Catastrophic Form and Planetary Realism." *New Literary History* 51 (2): 419–453.
George, James. 2006. *Ocean Roads*. Wellington: Huia Press.
Goin, Peter. 1991. *Nuclear Landscapes*. Baltimore, MD and London: Johns Hopkins University Press.
Grebowicz, Margret. 2021. "Ecology after Dark: Chernobyl's Wild Horses and the Traffic in Desire." *Minnesota Review* 96: 56–68.
Gusterson, Hugh. 1999. "Nuclear Weapons and the Other in the Western Imagination." *Cultural Anthropology* 14 (1): 111–143.
Hecht, Gabrielle. 2003. "Globalization Meets Frankenstein? Reflections on Terrorism, Nuclearity, and Global Technopolitical Discourse." *History and Technology* 19 (1): 1–8.
Hecht, Gabrielle. 2007. "A Cosmogram for Nuclear Things." *Isis* 98 (1): 100–108.
Hecht, Gabrielle. 2010. "The Power of Nuclear Things." *Technology and Culture* 51 (1): 1–30.
Hesse-Honegger, Cornelia, and Peter Wallimann. 2008. "Malformation of True Bug (Heteroptera): A Phenotype Field Study on the Possible Influence of Artificial Low-Level Radioactivity." *Chemistry and Biodiversity* 5 (4): 499–539.
Johnson, Rebecca L. 2015. *Chernobyl's Wild Kingdom: Life in the Dead Zone.* Minneapolis, MN: Twenty-First Century Books.
Keating, Fiona. 2017 "'Remarkable': Scientists amazed by thriving marine life at Bikini Atoll site where 23 atomic bombs were dropped." *The Independent*. July 15. www.independent.co.uk/news/science/fish-nuclear-weapons-bombs-sea-stanford-university-us-tests-hiroshima-a7842436.html.
Lander, Ned. 2011. *Dirt Cheap 30 Years On: The Story of Uranium Mining in Kakadu*. Northern Territory, Australia: Gundjeihmi Aboriginal Corporation in association with The Environment Centre.
Large, Timothy, dir. n.d. *Nuclear Wasteland*. Thomson Reuters Foundation. https://news.trust.org/item/20160817160135-pgb6h
Lee, Edmund W. J., and Shirley Ho. 2017. "Are Photographs Worth More Than a Thousand Words? Examining the Effects of Photographic-Textual and Textual-

Only Frames on Public Attitude toward Nuclear Energy and Nanotechnology." *Journalism and Mass Communication Quarterly* 95 (4): 948–970.

Marder, Michael, and Anaïs Tondeur. 2016. *The Chernobyl Herbarium: Fragments of an Exploded Consciousness*. London: Open University Press.

Marples, Megan, and Rachel Ramirez. 2022. "The Doomsday Clock Reveals How Close We Are to...Doom." CNN. January 20. Accessed March 30, 2022. https://edition.cnn.com/2022/01/20/world/doomsday-clock-2022-climate-scn/index

Masters, Dexter, and Katharine Way, eds. 1946. *One World Or None*. New York: McGraw-Hill.

Matashichi, Ōishi. 2011. *The Day the Sun Rose in the West: Bikini, the Lucky Dragon, and I*. Translated by Richard H. Minear. Honululu: University of Hawai'i Press.

Medvedev, Grigoriy. 1987. *Chernobyl Notebook*. eBook. Kindle Edition.

Morrison, Philip. 1946. "If the Bomb Gets Out of Hand." In *One World or None*, edited by Dexter Masters and Katharine Way, 1–6. New York: McGraw-Hill.

Nakazawa, Keiji. 2009. *Barefoot Gen: A Cartoon Story of Hiroshima*. Vol 10: *Never Give Up*. Translated by Project Gen. San Francisco, CA: Last Gasp.

Nancy, Jean-Luc. 2015. *After Fukushima: The Equivalence of Catastrophes*. Translated by Charlotte Mandel. New York: Fordham University Press.

Nash, Terre, dir. 1982. *If You Love This Planet*. National Film Board of Canada.

Nayar, Pramod K. 2023. *Nuclear Cultures: Irradiated Subjects, Aesthetics and Planetary Precarity*. New York: Routledge.

Oppenheimer, J. R. 1946. "The New Weapon: The Turn of the Screw." In *One World or None*, edited by Dexter Masters and Katharine Way, 22–25. New York: McGraw-Hill.

Prakash, Shri, dir. 1999. *Buddha Weeps in Jadugoda*. Bangalore: Pedestrian Pictures.

Ragan, Philip, dir. 1946. *One World or None*. Washington, DC: National Committee on Atomic Information and Philip Ragan Productions

Renck, Johan, dir. 2019. *Chernobyl*. New York: HBO.

Roy, Arundhati. 2002. *The Algebra of Infinite Justice*. New Delhi: Penguin.

Schnatz, Kerstin, Isabel Huber, and Stepanie Auth. dir. 2009. *Uranium: Is It a Country?* Berlin: Initiative Strahlendes Klima e.V.

Shields, Lora M., and Philip V. Wells. 1962. "Effects of Nuclear Testing on Desert Vegetation." *Science* 135 (3497): 38–40. New Series.

Sperling, Alison. 2020. "Radiating Exposure." In "Weathering: Cultures of Exposure", edited by Christoph F.E. Holzhey and Arnd Wedemeyer, special issue, *Cultural Enquiry* 17: 41–62.

Stephens, Lloyd, and Rosemary Barrett. 1978. "A Brief History of a 20th-Century Danger Sign." *Lawrence Berkeley National Laboratory*. LBNL Report #: LBL-7260. Accessed March 16, 2022. https://escholarship.org/uc/item/7cz9p0m6

Taylor, Bryan C., and Judith Hendry. 2008. "Insisting on Persisting: The Nuclear Rhetoric of "Stockpile Stewardship." *Rhetoric & Public Affairs* 11 (2): 303–334.

Thompson, Helen. 2014. "Chernobyl's Bugs: The Art And Science Of Life After Nuclear Fallout." *Smithsonian Magazine* 26 April. "http://www.smithsonianmag.com/"www.smithsonianmag.com/arts-culture/chernobyls-bugs-art-and-science-life-after-nuclear-fallout-180951 231/.

United Nations. "Doctrines of Dispossession." Accessed January 27, 2024. www.un.org/WCAR/e-kit/backgrounder4.htm

3 *Le Transperceneige* (1982) and the *Snowpiercers* (2013; 2020) as post-apocalyptic cli-fis

(Im)possible technologized habitats for the vulnerable posthuman Other

Cristina M. Gámez-Fernández

Introduction: post-apocalyptic cli-fi narratives as potential generators of human agency

This chapter will analyze three post-apocalyptic climate fiction narratives. First is *Tome 1: L'échappé* (*The Escape*), the first installment of the French graphic novel *Le Transperceneige*, authored by Jacques Lob, Benjamin Legrand, and Jean Marc Rochette ([1982] 2014). Second is Bong Joon-ho's film *Snowpiercer* (2013), offering a cinematic adaptation of this graphic novel over 30 years later. Finally, there is the first season of the TV series *Snowpiercer* (Friedman and Manson 2020) that reinterprets the two preceding narratives.[1] The assumption of this chapter is that these three post-apocalyptic stories dramatize the Janus-face nature of technology, which brings both development and progress and also climate change, so endangering future habitability on earth. The chapter aims to explore the historically situated responses of the three versions by addressing two pivotal questions. Firstly, are technological alternatives viable for sustaining life on our planet? Secondly, how do these technologized responses impact human vulnerability? To approach them, the chapter will compare the construction of plot and character in these post-apocalyptic stories. It will then examine how their geopolitical and socioeconomic contexts influence their depictions of climate change by considering their ideological and ecological anxieties in relation to (post-)colonial and neoliberal survival strategies. Subsequently, it will assess the technological features of each train and how such characteristics foster various social and political phenomena that potentially imperil the survival of humanity. Finally, the chapter will reflect on the material-discursive and visual construction of the trains as technologized ecosystems in an inhospitable nature/environment as well as on their entanglement with vulnerable posthumans.

The justification for selecting these three versions of the same story stems from Gregers Andersen's (2020) delineation of the productive

DOI: 10.4324/9781003383598-5

intersection of cultural analysis and cli-fi (2020), a term defined by Antonia Mehnert (2016) as "literature dealing explicitly with anthropogenic climate change" (4). Andersen contends that the proliferation of diverse cli-fi cultural products – including literary ones – underscores the promise of this genre for interdisciplinary research. Furthermore, Andersen observes that western cli-fi genres reflect a great "diversity of imaginaries" (2020, 7). Andersen traces the genesis of cli-fi to the early 1970s, noting that these narratives overlooked the true cause of climate change until the late 1970s, when its etiology was identified. However, a notable surge in western cli-fi production occurred in the late 1980s and the 1990s and culminated in a significant increase after 2010 (Andersen 2020, 7). The selected texts, therefore, exemplify the various cultural formats of western cli-fi, and showcase the historical evolution of the genre, as will be elucidated in the fourth section.

Examining the key features of post-apocalyptic cli-fi is pertinent for addressing this chapter's questions. Environmental degradation has gained prominence in literary and critical theory. An early example is Lawrence Buell's (2005) statement that "the power of story, image, and artistic performance and the resources of aesthetics, ethics and cultural imagination are crucial" to create a shared sense of environmental awareness (vi). However, Antonia Mehnert (2016) defined Dan Bloom's now widely spread abbreviation "cli-fi" only in 2016.[2] Axel Goodbody and Adeline Johns-Putra (2019) define cli-fi as "cultural work" which engages "with anthropogenic climate change, exploring the phenomenon not just in terms of setting, but with regard to psychological and social issues, combining fictional plots with meteorological facts, speculation on the future, and reflection on the human-nature relationship" (2). Regardless of their cultural format, cli-fi narratives vividly depict environmental changes as "explosive and spectacular" (Nixon 2011, 2) rather than slow or imperceptible (15).

The concept is subdivided into both disaster or calamity and apocalyptic or post-apocalyptic cli-fi. Justyna Poray-Wybranowska (2021) warns that, although there is an overlap between disaster fiction and apocalyptic fiction, "disaster and apocalypse [...] are not interchangeable in the environmental humanities" (44), because of apocalypse's religious origin, unlike that of disaster or catastrophe narratives. Defined by Laurence Buell as "the single most powerful master metaphor" (1995, 285) and by Greg Garrard (2023) as a "rhetorical" trope (102), apocalypse "has become an allegory, a rhetorical figure that enables us to speak of the present" (Ritzenhoff and Krewani 2016, xii). Early definitions of apocalyptic literature refer to a revelation about the end of history usually exposing a "struggle between good and evil" (Thompson 1997, 14). Kate Rigby (2015) explains apocalyptic narratives and their function, drawing on the

Greek concept of *kairos* in the Christian tradition, as "a moment of grave danger that also harbors liberating potentials" (19). When presented in post-apocalyptic cultural narratives, *kairos* provides an opportunity "to break […] with existing trends and tendencies" (19).

In contrast to these considerations of the power of apocalypse as a trope, however, Rigby cautions that post-apocalyptic renderings may not necessarily promote preventative climate change action since "it all depends, however, on how such calamities are framed" (2015, 10). Similarly, Amitav Ghosh (2016) coins the term "the great derangement" to describe contemporary artists' inability to represent the pressures of climate change in literature and the arts. For Ghosh, future human progeny may conclude that artists of our times employ "modes of concealment that prevent […] people from recognizing the reality of their plight" (11). Interestingly, post-apocalyptic cli-fi diverges from other cli-fi modalities. In this genre, the strained relationship between the imagination's power and the consequences of anthropogenic climate change takes center stage because artists imagine the fatal consequences of climate change. Therefore, the capacity of post-apocalyptic cli-fi to represent such spectacular and explosive events (Nixon 2011) enables perception of the otherwise invisible prospects of anthropogenic agency in endangering future planetary habitability.

Astrid Bracke's (2018) commentary on the potential of narrating anthropogenic climate change resonates with the historical situatedness of the three cli-fi narratives analyzed in this chapter: "climate crisis narratives […] also function as frameworks through which people interpret the world around them" (10). Moreover, in contrast to her claim that "apocalyptic narratives […] present a world that is falling apart, giving viewers and readers a sense that at least the 'real' world is fine" (Bracke 2018, 25), the selected texts can be productively analyzed through the lens of what Erin James (2015) describes as the principle of minimal departure. This theory refers to the study of fiction in which readers are engaged in "considering and integrating both the world that is in the narrative and the world that is not" (xi). Through their portrayal of dystopian climate change scenarios, initially perceived as improbable, these three post-apocalyptic cli-fis compel reflection on their underlying historical and ideological contexts.

Emplotment and character design across three media

Based on the French graphic novel *Le Transperceneige*, the release in 2013 of the movie *Snowpiercer* aroused great interest on social media and the Internet about the original source material. *Le Transperceneige*'s writer Jacques Lob had commenced a collaboration on the project in the 1970s with a famous cartoonist pen-named Alexis. However, the graphic novel came to a halt on Alexis' death in 1977. *Tome 1: L'échappé*, the first installment of *Le Transperceneige*, was finally published by the Franco-Belgian publisher Casterman in 1982 when artist Jean-Marc Rochette became illustrator. After Lob died in 1990, *L'échappé* was continued by writer Benjamin Legrand working with Rochette in two subsequent installments, entitled *L'arpenteur* (1999) and *La traversée* (2000). Influenced by the 2013 movie, the French graphic novel was published in English translation in 2014 by Titan Comics with the three French volumes condensed into two, *The Escape* and *The Explorers*, as the latter included *The Crossing*. In turn, the international attention generated by the movie encouraged completion of the French series of the graphic novel in 2015, when scriptwriter Olivier Bocquet and Rochette published *Terminus*, translated a year later into English. The saga gained momentum with the 2020 eponymous TV series – produced by TNT and available through Netflix – which released its third season in 2022. Hence, the intricate interplay between the French graphic novel, the movie, and the TV series highlights how this post-apocalyptic cli-fi story has been adapted across transnational, transcultural, and temporal boundaries.

The storyline iterations in these three media depict a train relentlessly moving after the earth has frozen, facilitating the survival of the last remnants of human civilization and other species, primarily for human subsistence. This recurring topos is articulated differently in terms of plot and character design within each medium or cultural product. In the first installment of the graphic novel, *L'échappé*, a man called Proloff manages a miraculous escape from the rear car of the train, where an indeterminate number of people is confined. Military forces detain him and decide to interrogate him in the train's headquarters located at the opposite end of the convoy, disregarding the doctor's recommendation to isolate him due to potential public health risks. Adeline Belleau, a member of the train's humanitarian group advocating for the assimilation of those from the tail's car into the train, attempts to assist him but she is also detained. *L'échappé* portrays their forced journey to the front through the train's various cars, revealing sections inhabited by different social groups with sundry degrees of wealth that offer a panoramic view of societal degeneration and decline. Proloff and Belleau finally meet the authorities at the front, Colonel Krimson and the president, who explain to them that the train's engine is slowing down due to excessive weight and request their assistance in relocating the tail car's passengers to the train before disconnecting those wagons. However, Proloff uncovers the true plan – which is to disconnect the tail car with its inhabitants and the humanitarian activists inside – and he manages to warn them about it. Meanwhile, an unknown disease breaks out onboard, allegedly spread by Proloff, who had resorted to cannibalism in the tail car to survive. Belleau dies accidentally at Proloff's hands, and he is eventually rescued by Alec Forrester, the engineer responsible for the train's engine. As Forrester falls ill, he asks Proloff to take care of the engine. With the virus having spread and killed everyone on board, Proloff is left alone in the engine car.

In the movie, set 17 years after the earth's freezing, a train circumnavigates the globe propelled by a perpetual motion engine devised by a wealthy Mr. Wilford, who resides in it. Wilford's minion, Minister Mason, and her soldiers subjugate the tail car occupants through rationing their food, exerting population control, and enforcing disciplinary measures, citing their unauthorized boarding of the train as justification. Periodically, they also take children away after measuring them for reasons unknown. In response, the marginalized crowd conceives a plan to reach the engine at the front, wrest control of the train, and redistribute wealth more equitably. The uprising's leader, Curtis Everett, along with his team, liberates the security engineer Namgoong Minsoo and his daughter Yona, who promise to unlock the sealed doors between cars. During their mission, Curtis loses his team members, eventually arriving at the engine of Mr. Wilford, who reveals that he himself had instigated the rebellion as a means of population control. Wilford offers his position to Curtis but warns him about the engine's limitations, revealing then that the "measured" children are used as surrogates for essential, degraded components of the machine.

In the TV series, the tail cars' denizens devise a plan to revolt against their meagre living conditions that the authorities justified because they lacked tickets to board the train. They are frequently summoned from the tail to provide specific services, such as playing the violin in a social gathering held in the first-class cars. Andre Layton, one of the leaders of the tail's inhabitants and a former homicide detective, is unexpectedly called by Melanie Cavill from the hospitality department to assist her in solving a crime. Despite the imminent rebellion, Layton persuades them to postpone it temporarily, so he can devise a more effective, strategic revolution.

The three plots and characters' roles effectively provide space for reflection on how historically evolving sociopolitical and economic processes impact on the depiction of vulnerable and precarious Others and their attempts to survive. This will be further explored in the two subsequent sections.

Global warming in three media: from (post)colonial to neoliberal anxieties

According to Sara Keller (2020), "the positive reception of *Snowpiercer* [...] and other films like it suggests that they successfully tell us something about ourselves that speaks to our anxieties" (191). Studying these three cli-fi responses to the same dystopian catastrophe sheds light on the unique (post)colonial and neoliberal anxieties that emerge in each, revealing diverse historical roots, social dynamics of oppression, and ethical concerns, in addition to their idiosyncratic aesthetic conformity. According to Poray-Wybranowska (2021), there is a strong linkage between present (or future) vulnerability to climate change and "the historical roots of present-day ecological crises" (13) that postcolonial studies concern itself with. For Graham Huggan (2004), "current ecological crises cannot be separated from historical legacies of imperialistic exploitation and authoritarian abuse" (702). Thus, the power struggle in the three narratives recasts anew (post)colonial and neoliberal anxieties when depicting complex, usually abstract, phenomena associated with hegemonic power contestations in the lives of individual characters.

Despite their specificities, some commonalities characterize the three cli-fis. Firstly, they represent "engineered atmospheres" as "places of incarceration from which one or several protagonists long(s) to escape" (Andersen 2020, 109). Even though the characters manage to evade an impending frozen apocalypse, albeit with different degrees of success in each story, they find themselves ensnared in the same oppressive economic, social, and cultural practices that existed before the catastrophe forced them to this mechanistic refuge. Locked within the trains' walls, the privileged occupants in each narrative repeatedly enact a caged version of their previous opulent lifestyles, while the others, confined to ceaseless struggle for recognition,[3] provide a post apocalyptic image of contemporary global maladies. These populations exemplify various modes of precarity, which according to Judith Butler (2009), is "a politically induced condition" in which "certain populations suffer from failing social and economic networks of support and become differentially exposed to injury, violence, and death" (25). The three cli-fi narratives adapt previously known forms of (post)colonialism and neoliberalism to focus on survival strategies aimed at overcoming these autocratic socio-political regimes. Another common element in these cli-fi stories is a power dynamics enforced by political regimes that either re-name nonviolent acts as violent manifestations – or the opposite – to safeguard their capacity for employing systemic violence, their social, and legal justification of such use, and their narrative monopoly over it (Butler 2020). A remarkable example is found in the French graphic novel *L'échappé*: during Adeline Belleau's initial visit to Proloff, he enquires if, like him, other passengers from the tail cars ever made it to the train. She mentions that rumors only circulated about "The Wild Assault", a significant historical event within the train. To such a reference, Proloff replies in surprise: "Are you talking about 'The Massacre'?" (Lob, Legrand, and Rochette [1982] 2014, 15),[4] thus showing how the same event is differently referenced, according to the social group and perspective involved.

In the aftermath of these eco-catastrophes, these three post-apocalyptic texts depict different ethico-onto-epistemological (Barad 2007) and ecosocial configurations of "transnational, transspecies and trans-corporeal connectivities" (Rigby 2015, 77). Thus, these cli-fis narrativize distinct survival strategies tailored to their respective contexts, rendering these disasters as "material-discursive processes" (Rigby 2015, 15). They eschew focusing solely on "the impact of the disaster" such as the quick freezing of the globe, centering instead on "what happens afterward" (Rigby 2015, 14) once all forms of life are extinct on earth. As shown in each version, such configurations range from instances of "opportunistic looting and apocalyptic hedonism through to pragmatic self-organization on the part of communities and selfless kindness among some individuals" (Rigby 2015, 77). In the graphic novel, the humanitarian emergency in the tail cars is not portrayed, but the narrative decisions depicted in the two audiovisual adaptations of the novel constitute an agentic response in addressing their harsh conditions, based in communitarian forms of grassroot organization. Additionally, the emplotment and motivations of the protagonists in the different versions point towards the specific ideological underpinnings that uniquely characterize each of their contemporary zeitgeists. Proloff is depicted in *L'échappé* as a loner who decides to risk his life outside the train to escape the tail cars without any connection to other inhabitants of the tail. In Joon-ho's film, Curtis leads an organized revolt to redeem himself from his past murders and acts of cannibalism. Lastly, in the TV series, Layton is portrayed as a committed democrat and advocate for social and economic justice, engaging in constant dialogue with various social groups to achieve his objectives of social justice and egalitarianism.

These cli-fi narratives can be read through the theoretical lens of Saskia Sassen (2014), who argues that the traditional tools to measure poverty and injustice fail to describe the consequences of neoliberal practices. According to Sassen, imperial practices were geopolitically located around the traditional binary of west and east. Both the graphic novel and the movie revolve around such customary conceptualizations as the consequences of authoritarian governments. Hence, as an allegory of French politics and the social unrest of its era, the graphic novel *L´échappé* depicts a clear-cut distinction between different social classes and a political system that invokes historically traditional powers in France of the 1970s and 1980s: military forces that follow the president's order, civilian authorities in the form of a train conductor, and the church, which attributes sacred powers to the engine. In addition, the opening chapter of the French graphic novel is titled "the rolling ghetto", the term "ghetto" being widely used in France in the late 20th century to refer to "areas where rampant poverty and growing violence were a vivid proof of the demise of the French social model" (Simon 1998, 42). The public disturbances shown aboard demanding that Adeline Belleau be freed, mirror the demonstrations and strikes against rising unemployment and the riots associated with youth of North African descent in 1979 and 1981 that finally helped François Mitterrand come to office. In Joon-ho's *Snowpiercer* (2013), these roles epitomize extreme 20th-century ideologies, strongly based on the two world wars and the rise of Nazism in Europe. The train's political system stratifies its passengers into only two classes: the oppressed, who are relegated to the squalid conditions of the windowless tail cars, and the elite, who inhabit the luxurious front cars.

In contrast to traditional representations of poverty and inequality illustrated by the French graphic novel and the film, the TV series can be analyzed through the lens of Sassen's term "predatory formation", an organizational entity consisting of "a mix of elites and systemic capacities" (2014, 13) that reach and remain at the top of the system. The TV series vividly portrays social struggles in four distinct social groups of passengers, paralleling a microcosm of the current global world order: the first class, representing the top one percent of the global population, which amasses 82 percent of the world's wealth (Oxfam 2023); the second class, comprising professionals and technicians whose role is essential for the train's operation and the provision of specific commodities; the third class, encompassing the proletariat; and the stowaways in the tail, emblematic of the masses of displaced, unemployed, disenfranchised, and refugee populations worldwide. The TV series features an image of 21st-century diversity among the tail's dwellers, who speak many languages, and have mixed cultural, ethnic, racial, and religious backgrounds. Ultimately, predatory formations result in "expulsions", i.e., being forced out "from professional livelihood, from living space, even from the very biosphere that makes life possible" (Friedman and Manson 2020, back cover). Predatory formations breed a new form of violence derived from the creation of a complex transnational system that lacks any identifiable or legally liable organizations. Such anonymity is illustrated in the TV series when, at the end of episode one, it is revealed that Mr. Wilford, whose name has been frequently mentioned and whose voice been heard over the train speakers, is a fabrication. A taste of the logics of expulsion appears in the requirement of a ticket for entry or a special invitation from this nonexistent character. Against the post-apocalyptic backdrop, the TV show exposes the forms of violence to which the "precariat class" (Standing 2011) is subject while they struggle to survive at the tail end of the train.

Resonating with Buell's, Rigby's, and Ghosh's references to the power of the imagination in literature and art concerned with climate change, Fredric Jameson's (2003) early warning about the entrenchment of capitalistic practices is also relevant to these media adaptations of *Le Transperceneige*: "it is easier to imagine the end of the world than to imagine the end of capitalism. We can now revise that and witness the attempt to imagine capitalism by way of imagining the end of the world" (76). Along the same lines, when exploring the historical situatedness of each post-apocalyptic cli-fi narrative, Jason W. Moore's (2016) conceptualization of the "Capitalocene" is particularly relevant in describing how capitalism's relentless pursuit of growth and profit becomes the source of a global degradation that perpetuates forms of (post)colonial and neoliberal injustice. Thus, the term is informed by the historical, geopolitical, socioeconomic, scientific, cultural, and ecological intertwined responses to capitalistic dynamics. Moore's criticism of a reductionist view of climate change is echoed by Mehnert: "climate change is considered an issue of concern solely for natural sciences and engineering practices. It is thus turned into a problem that is assumed to be solved best with a scientific-technological fix" (2016, 214),[5] thus implying that the real cause of planetary degradation remains willfully ignored. Hence, as the next section will demonstrate, the three narratives challenge scientific-technological fixes rooted in capitalist logics. These solutions not only prove short-term remedies but also perpetuate the degradation of the *oikos*, a Greek term referring to the earth as the sole habitat for life.

Climate change in three media: feasibility and equity in post-apocalyptic technologized survival

Andrew Milner and J.R. Burgmann (2020) describe climate change in science fiction by identifying "three main tropes: the new ice age, the burning world and the drowned world; or, more succinctly, ice, fire and flood" (3). The authors' comment that "the most likely future climate change was widely anticipated to be a return to the ice age" (14) is shown in these three post-apocalyptic cli-fis. When discussing the history of comics and graphic novels dealing with climate change, Milner and Burgmann set aside *L'échappé* and *L'arpenteur* (the first two sequels), "which take anthropogenic global cooling as their mise en scène" (176). They contrast Joon-ho's *Snowpiercer* with the French graphic novel, asserting that *Le Transperceneige* employs the ice age trope by "strongly hint[ing] that the world has been destroyed by fiercesome new climate weapons", yet lacks "much evidence of interest in or knowledge of current scientific speculation about the possibility or likelihood of a new real-world ice age" (19–20).

The three stories instantiate narratives of what Rigby describes as "Faustian geo-engineering schemes" (2015, 16) to counter global warming. In all three narratives the "hermeneutics of suspicion" (Andersen 2020, 12) is strategically employed by such predatory formations (Sassen 2014) to divert attention from anthropogenic global warming. In the graphic novel, for example, Proloff and the librarian represent the debate about the causes of the earth's icing up. Epitomizing the erudite scholar, the librarian argues that the origin of the global freeze is unknown – "we never found out what was the cause of the catastrophe" (Lob, Legrand, and Rochette [1982] 2014, 63) – whereas Proloff insists that it was caused by a nuclear bomb, stating "it just happened when the war was declared" (63). These two characters dramatize a larger dysfunction whereby epistemological uncertainty and inconsistency generate the blur between fact and belief, "science and politics, and truth and ideology" (Andersen 2020, 12).[6] Given its publication date, the nuclear arms race of the Cold War naturally remains in the background of *L'échappé*.[7] In contrast, both audiovisual productions describe the cause of the ice age in detail as a failed geoengineering experiment, consisting in the release of CW-7, a coolant chemical, into the atmosphere to counteract global warming effects. This justification paradoxically resonates with Rigby's description of "natural disaster" as a "modern myth" (2015, 4), a misnomer that evades human responsibility for ecological deterioration. Instead, a natural Other/order with a violent response is envisioned, triggering ecophobic (Estok 2009) discourses identified by Susanna Hoffman and Anthony Oliver-Smith (1999) as "the angry earth". Moreover, these discourses perpetuate and reinforce the anonymity and lack of accountability characteristic of neoliberal predatory formations (Sassen 2014) as depicted in the TV series.

The three narratives reflect a dynamic widely addressed by climate change scholars, wherein certain populations are more vulnerable than others to climate change due to factors like age, sex, gender, race, ethnicity, Indigencity, or economic status (Dow, Kasperson, and Bohn 2006; Barnett 2006). They display human groups deprived not only of a livable habitat but also of "the security provided by societal institutions" (Andersen 2020, 11). Such "social collapse" is part and parcel of the acceleration of "social disintegration and interhuman violence on a very large scale" (11) caused by global warming and the failed attempts to control it. These post-apocalyptic narratives comply with what Christian Parenti (2011) calls "catastrophic convergence", i.e., that climate change effects "compound and amplify" (7) preexisting problems related to poverty and violence. Andersen identifes this phenomenon as "two forms of emotional and physical vulnerability" (2020, 11), one referring to their exposure to violence inflicted by other humans and the Other to their deadly exposure to gelid temperatures. Examples include punishment by freezing limbs, as happens in both audiovisual media, and the common trope of confining large masses of people to the tail cars without basic necessities.

Additionally, the three cli-fis narrativize Malthusian logics of population management onboard. *Le Transperceneige* and Joon-ho's *Snowpiercer* and in different ways "attempt [...] to envision the biopolitics of the future" (Andersen and Nielsen 2018, 618): the graphic novel articulates the governmental decision to remove the excess weight of disposable humans, and the film degrades humans to the status of breeding or producing children-components. Both explore thanatopolitical practices that regulate overpopulation "biopolitically and anatomo-politically" (Andersen 2020, 110) in the face of post-apocalyptic ecological collapse and resource paucity. In addressing thanatopolitical issues in Joon-ho's *Snowpiercer*, Gerry Canavan (2014) claims that despite the train's advanced technology, it ultimately becomes a prison for its inhabitants, highlighting the consequences of unchecked technological hubris, where a decadent society is on the brink of collapse, unable to avoid death. In turn, the TV series narrativizes how certain people are put into stasis – a form of hibernation. However, despite such differences, the three stories ironically suggest that the sacrifices made to survive "remain solely the burden of the tail-section subalterns" (Schwetman 2021, 75). Rather than providing answers, the engineered material-discursive construction of post-apocalyptic livable worlds poses the question of "social design", specifically "how many human and nonhuman beings can be allowed to take up the space of the engineered atmosphere without risking its ecological equilibrium" (Andersen 2020, 135) and how such decision making affects their vulnerability. This will be explored in the next section.

The material-discursive construction of trains as entangled with vulnerable posthumans

According to John D. Schwetman (2021), the train is "a literary icon of the Industrial Revolution" whose presence can be traced "through the Information Age and, into the dystopian future of [Joon-ho's] the *Snowpiercer*" (67). The material-discursive (Barad 2007) construction of the three trains and their engines exacerbates the power dynamics whereby the Anthropocentric worldview has historically justified western social, political, and economic systems. The double valence of the train hinges on "human development and progress that is simultaneously a harbinger of humanity's destruction" (Schwetman 2021, 67). Moreover, such material-discursive construction directly impacts the depiction of the humans occupying the three post-apocalyptic trains. The figure of the train encapsulates two productive symbols that also contribute to shaping its precarious existence alongside its inhabitants. First, a modernized, mechanical version of the ouroboros, namely the Norse myth of the Jörmundgandr, the world serpent that grew so large that it encircled the globe and ate its own tail. Notions traditionally evoked by the ouroboros, such as the cyclical nature of things or the eternal return, ultimately symbolize the continuity of life, since the creature's rebirth suggests that cycles recur eternally. However, the ouroboros takes on different roles in each cultural version. In the graphic novel, the train with its 1,001 cars continues to roll on without any survivors except Proloff, suggesting the end of humanity.[8] In contrast, Joon-ho's film adaptation shows the train's destruction, including those responsible for political oppression, offering a fresh start for humanity as the earth gradually thaws. Unlike the other two cli-fis, the TV *Snowpiercer* depicts an electronically and digitally driven ouroboros, closely resembling the dynamics of the Fourth Industrial Revolution, which captures the perpetually shifting and kaleidoscopic struggle against the neoliberal forms of oppression denounced by Moore's Capitalocene.

Second, the train symbolizes an ever-moving technological version of Noah's ark. The three cli-fis offer a modern example of Mary Shelley's ([1826] 2004) *Last Man*, in which a boat populated by a man and a dog navigates coastlines in search of survivors from a world pandemic. Similarly, Andreas Malm and Alf Hornborg (2014) find the concept of the boat particularly useful as they disagree with Dipesh Chakrabarty's (2009) proposal that the Anthropocene will provide no lifeboats for the wealthy (221). As these three dystopian renderings demonstrate, "there *will* be lifeboats for the rich and privileged" (Malm and Hornborg 2014, 5; emphasis in original). Unsurprisingly, *L'échappé* is the only version to clearly invoke Noah's ark in the repeated mention of the number of carriages, echoing the specificities provided by God to Noah to build the ark (Schwetman 2021, 75–77).

Similarly, the qualities and requirements of the train's engines condense the ideological core propelling these narratives, offering new insights into the vulnerable posthuman. For Rosi Braidotti (2013), the notion of the posthuman blurs the boundaries of the human and nonhuman and defines subjects as mutable and interdependent with other human and nonhuman bodies, material forms, and machines. The three narratives subordinate the right to exist of disenfranchised individuals, who are subject to the demands of the machine, converting their already precarious existence into a state of being "less than human" (15). In the graphic novel, communal affiliations progressively deteriorate as religious devotion to the holy engine emerges, causing people to worship the machine rather than the beings it safeguards. In the movie, the engine's need for specific pieces justifies the existence of the tail car dwellers as only suppliers of children who become "cyborg*ed*" – as biological components of the engine. Since "in a large, complex system, the failure of one component can produce system wide failure" (Schwetman 2021, 79), this significantly dissolves the boundaries of both machine and humans, in this way depicting a mechanistic understanding of human life. In the film, Minister Mason lectures the confined masses in the rear cars about order, the organizing principle that sustains life in the train: "We must each of us occupy... our par-tic-u-lar, pre-ordained position! In the beginning, order was prescribed by your ticket [...]. Eternal order is prescribed by the Sacred Engine" (Joon-ho 2013, 00:17:19–00:17:21; 00:17:55–00:18:03). In the TV series, inter-class dialogue to seek social justice is contingent upon technological adjustments required by the train. The technologized version of the snake and the ark is undergirded by the idea of the porousness integral to posthuman nature. On one level is the fact that the protection against the extreme, cold temperatures of the earth can be removed at will to punish those at fault or make humans and other species vulnerable to unforeseeable eventualities, such as when the last cows freeze in an avalanche that breaks the car's windows in the TV *Snowpiercer*. On another level is the apparently fixed class rank designated for the different train cars. This serves as a metaphor for the social class struggle depicted in the TV *Snowpiercer*, where Melanie uses class mobility as an incentive for those who complete specific tasks.

The entanglement of the material-discursive construction of the trains with the representation of vulnerable posthumans and their precarious survival strategies is portrayed in the three cli-fi narratives by the (audio) visual depictions of each train. *L'échappé* showcases the most compelling pictorial representation of both the ouroboros motif and religious undertones reminiscent of Noah's ark. Its black-and-white panels with multiple aerial views of the train emphasize the ceaselessly rolling metallic serpent and its extraordinary length. Particularly striking is the final image, depicting the spectral navigation of the black ark-train with all corpses aboard, set against a white background. While the three narratives present articulate renderings of the train cars' interiors to contrast poverty and luxury onboard, Joon-ho's film veers towards the absurd with its portrayal of extremes, as exemplified by the nightclub car panning shots. As sites of confinement, the different narratives utilize images and cinematographic techniques to evoke the atmosphere of constraint and control. Individuals in *L'échappé* are subject to procedures such as isolation and forced head shaving, while the TV series devises suspension chambers to induce stasis, visually akin to a morgue. Furthermore, the three trains graphically reinforce the precarious quality of their technological structures. In *L'échappé*, the scientific argument is simplified to a matter of excess weight, leaving unanswered the question of whether the train, laden with corpses, will ever come to a halt. In the film, the stark contrast between the tail cars and the rest is illustrated primarily through shots of gloomy, claustrophobic cars juxtaposed with brightly colorful and spacious ones, as well as the unexpected revelation of hidden spaces. Examples are the discovery of mechanically operated holes used to freeze people's limbs or minuscule cavities where children-pieces enable the train's movement. Finally, in the TV series, the challenges posed by the perpetual-motion engine are repeatedly depicted through high-end, colorful screens, and sophisticated digital control panels. Despite its polished appearance, the train continues to trigger alarm messages of danger that foreground social struggles.

The constant risk of the train engines' failure in the three cli-fis suggests that humans continue to create fail-safes in their delusory, futile attempt to control nature's responses to anthropogenic change and protect themselves. Moreover, the three trains are subject to conflicting visions of a technologized version of terraforming within their cars, precariously enabling life in the three postapocalyptic scenarios. In this context, the trains transform into a previously unexplored technologized space essential for human survival. As passengers board them, they transmute from being symbols of consumerism into being custodians of civilization. Subsequently, the trains become either a non-place (Augé 1995), where disenfranchised individuals remain anonymous without lasting social connections, or an anthropological place, imbued with social and cultural meaning for the privileged. The trains precariously supplant Mother Earth – the engine is addressed as female in the three cli-fis – because they embody flimsy technological ecosystems. Historical struggles on earth are perpetuated within the trains' wagons, further complicating such power contestations due to the reduced space available, resource scarcity, and impending risk of extinction. Each of the three cli-fis thus enacts a fabula showcasing the delusional vision of human capacity to control natural ecosystems, which remain permeable and subject to unforeseen contingencies.

Conclusion: the tension between cyclicality and *kairos* in post-apocalyptic media

The analysis of *Tome 1: L'échappé* from the French graphic novel *Le Transperceneige*, Joon-ho's film *Snowpiercer*, and the eponymous TV series has provided key insights on the two questions posed at the beginning of this chapter. To the question, "are technological alternatives viable for sustaining life on our planet?", the study of these three post-apocalyptic cli-fis suggests that technological alternatives for sustaining life on our planet strongly depend on sundry contextual factors in which the three media were created. In this sense, the cli-fis resonate with the anxieties of the contexts in which they were produced. As such, *L'échappé* resorts to a complete annihilation of civilization caused solely by human, not technological failure. In contrast, the film advocates the destruction of both human and technological systems when only two survivors – a young woman and a boy – leave the train wreck and walk into a resilient, warming natural landscape. The first season of the TV series, in turn, concludes by revealing the existence of another train circumnavigating the globe while aboard mayhem governs after the rebellion of the tail section. In the three media, these alleged technologized "havens" either perpetuate pre-apocalyptic forms of oppression or exacerbate them to the point where the survival of humanity is gravely compromised or annihilated.

The answer to the question "how do these technologized responses impact human vulnerability?" is underpinned by the systemic structure of such technologies. In each of the three post-apocalyptic cli-fis, oppressive systems exist that guarantee the luxurious lifestyles of certain populations at the expense of intensifying the precarity of others. Moreover, only the precarious and vulnerable disenfranchised populations in each of the three trains become posthuman others, permeable to the technologized space that shelters them, with human-machine boundaries dissolved. Similarly, the technological ecosystem is anthropomorphized as a mother, revered as a deity, or prioritized at the expense of multiple human lives. Each of the narratives prompts audiences and readers to acknowledge the interconnectedness of planetary survival and human existence, while reflecting on the underlying ideologies of civilization. They help to explore the tension between the cyclic nature or perpetual existence of oppressive systems, which cannibalize vulnerable posthuman others just as they exhaust the natural resources of our *oikos*, and the inherent potential of *kairos* as a tipping point in the face of total annihilation.

Notes

1 For reasons of space, I focus exclusively on the initial installment of the French graphic novel and the inaugural season of the TV series for this analysis.
2 See Dan Bloom's articles on "cli-fi": https://www.cli-fi.net/index
3 Note, given the role of the trains, Carrigan's (2011) connections between tourism – particularly how the tourist industry furthers historical practices of colonialism – the history of colonialism, and ecological degradation.
4 Quotations from the graphic novel belong to the English translation published by Titan Comics.
5 As will be described in the following section, two "scientific-technological fix[es]" (Mehnert 2016, 214) can be easily identified in the three narratives: first, the desperate actions devised to control global warming both through climate bombs in the graphic novel and the chemical release to the atmosphere in the film and TV series, and second, the metal habitat represented by the trains.
6 The investment in the creation of a culture of doubt to guarantee inaction is acknowledged by Nixon (2011, 39) as a feature of western practices that foster slow violence.
7 Clint Jones (2020) has cleverly commented on the inconsistencies of the graphic novel over the consequences of an ice age caused by nuclear blasts.
8 The following installments in *Le Transperceneige* prove otherwise.

References

Andersen, Gregers, and Esben Nielsen. 2018. "Biopolitics in the Anthropocene: On the Invention of Future Biopolitics in *Snowpiercer*, *Elysium*, and *Interstellar*." *The Journal of Popular Culture* 51 (3): 615–634.

Andersen, Gregers. 2020. *Climate Fiction and Cultural Analysis: A New Perspective on Life in the Anthropocene*. London and New York: Routledge.

Augé, Marc. 1995. *Non-Places: Introduction to an Anthropology of Supermodernity*. Translated by John Howe. London: Verso Books.

Barad, Karen. 2007. *Meeting the Universe Halfway*. Durham, NC: Duke University Press.

Barnett, Jon. 2006. "Climate Change, Insecurity, Injustice." In *Fairness in Adaptation to Climate Change*, edited by W. Neil Adger, Jouni Paavola, Saleemul Huq, and M. J. Mace, 115–130. Cambridge and London: MIT Press.

Bracke, Astrid. 2018. *Climate Crisis and the 21st-Century British Novel*. London and New York: Bloomsbury Academic.

Braidotti, Rosi. 2013. *The Posthuman*. Cambridge and Malden: Polity.

Buell, Lawrence. 1995. *The Environmental Imagination: Thoreau, Nature Writing, and the Formation of American Culture*. Cambridge, MA: Harvard University Press.

Buell, Lawrence. 2005. *The Future of Environmental Criticism: Environmental Crisis and Literary Imagination*. Malden: Blackwell Publishing.

Butler, Judith. 2009. *Frames of War: When is Life Grievable?* London and New York: Verso.

Butler, Judith. 2020. *The Force of Nonviolence: An Ethico-Political Bind*. London and New York: Verso.

Canavan, Gerry. 2014. "'If the Engine Ever Stops, We'd All Die': *Snowpiercer* and Necrofuturism." *Paradoxa* 26: 1–26.

Carrigan, Anthony. 2011. *Postcolonial Tourism: Literature, Culture, and Environment*. London and New York: Routledge.

Chakrabarty, Dipesh. 2009. "The Climate of History: Four Theses." *Critical Enquiry* 35 (2): 197–222. http://www.jstor.org/stable/10.1086/596640

Dow, Kristin, Roger E. Kasperson, and Maria Bohn. 2006. "Exploring the Social Justice Implications of Adaptation and Vulnerability." In *Fairness in Adaptation to Climate Change*, edited by W. Neil Adger, Jouni Paavola, Saleemul Huq, and Mary J. Mace, 79–129. Cambridge, MA and London: MIT Press.

Estok, Simon C. 2009. "Theorizing in a Space of Ambivalent Openness: Ecocriticism and Ecophobia." *ISLE* 15 (2): 203–225. https://doi.org/10.1093/isle/isp010

Friedman, Josh, and Graeme Manson, dir. 2020. *Snowpiercer*. Canada, TNT.

Garrard, Greg. 2023. *Ecocriticism*. 3rd ed. London and New York: Routledge.

Ghosh, Amitav. 2016. *The Great Derangement. Climate Change and the Unthinkable*. Chicago, IL and London: Chicago University Press.

Goodbody, Axel, and Adeline Johns-Putra. 2019. "Introduction." In *Cli-Fi: A Companion*, edited by Axel Goodbody and Adeline Johns-Putra, 1–19. Oxford: Peter Lang.

Hoffman, Susanna M., and Anthony Oliver-Smith, eds. 1999. *The Angry Earth: Disaster in Anthropological Perspective*. London and New York: Routledge.

Huggan, Graham. 2004. "Greening Postcolonialism: Ecocritical Perspectives." *MFS: Modern Fiction Studies* 50 (3): 701–733. www.jstor.org/stable/26286314
James, Erin. 2015. *The Storyworld Accord: Econarratology and Postcolonial Narratives*. Lincoln, NE and London: University of Nebraska Press.
Jameson, Fredric. 2003. "Future City." *New Left Review* 21: 65–79. https://newleftreview.org/issues/ii21/articles/fredric-jameson-future-city.pdf
Jones, Clint. 2020. *Apocalyptic Ecology in the Graphic Novel: Life and the Environment after Societal Collapse*. Jefferson, NC: McFarland.
Joon-ho, Bong, dir. 2013. *Snowpiercer*. Korea: Moho Films and Opus Picture.
Keller, Sarah. 2020. *Anxious Cinephilia: Pleasure and Peril at the Movies*. New York: Columbia University Press.
Lob, Jacques, Benjamin Legrand, and Jean-Marc Rochette. [1982] 2014. *Snowpiercer: The Escape*. London: Titan Comics.
Malm, Andreas, and Alf Hornborg. 2014. "The Geology of Mankind? A Critique of the Anthropocene Narrative." *The Anthropocene Review* 1 (1): 62–69. https://doi.org//10.1177/2053019613516291
Mehnert, Antonia. 2016. *Climate Change Fictions: Representations of Global Warming in American Literature*. Cham: Palgrave Macmillan.
Milner, Andrew, and J. R. Burgmann. 2020. *Science Fiction and Climate Change: A Sociological Approach*. Liverpool: Liverpool University Press.
Moore, Jason W., ed. 2016. *Anthropocene or Capitalocene? Nature, History, and the Crisis of Capitalism*. Oakland, CA: PM Press.
Nixon, Rob. 2011. *Slow Violence and the Environmentalism of the Poor*. Cambridge, MA: Harvard University Press.
Oxfam Press Release. 2023. "Richest 1% Bag Nearly Twice as Much Wealth as the Rest of the World Put Together Over the Past Two Years." January 16. https://www.oxfam.org/en/press-releases/richest-1-bag-nearly-twice-much-wealth-rest-world-put-together-over-past-two-years
Parenti, Christian. 2011. *Tropic of Chaos: Climate Change and the New Geography of Violence*. New York: Bold Type Books.
Poray-Wybranowska, Justyna. 2021. *Climate Change, Ecological Catastrophe, and the Contemporary Postcolonial Novel*. London and New York: Routledge.
Rigby, Kate. 2015. *Dancing with Disaster. Environmental Histories, Narratives and Ethics for Perilous Times*. Charlottesville, VA and London: Virginia University Press.
Ritzenhoff, Karen A., and Angela Krewani. 2016. "Introduction." In *The Apocalypse in Film: Dystopias, Disasters, and Other Visions about the End of the World*, edited by Karen A. Ritzenhoff and Angela Krewani, i–xxii. Lanham, MD: Rowman and Littlefield.
Sassen, Saskia. 2014. *Expulsions: Brutality and Complexity in the Global Economy*. Cambridge, MA: The Belknap Press.
Schwetman, John D. 2021. "A Genealogy of Apocalyptic Trains: *Snowpiercer* (2013) and its Precursors in the Transnational Literature of Transport." In *Transnational Railway Cultures: Trains in Music, Literature, Film, and Visual Art*, edited by Benjamin Fraser and Steven Spalding, 64–83. New York and Oxford: Berghahn.

Shelley, Mary. [1826] 2004. *The Last Man*. Introduction and Notes by Pamela Bickley. Ware: Wordsworth Editions.

Simon, Patrick. 1998. "Ghettos, Immigrants, and Integration of the French Dilemma." "Segregation, Concentration and Ghetto Formation", special issue, *Netherlands Journal of Housing and the Built Environment* 13 (1): 41–61. https://www.jstor.org/stable/41107731

Standing, Guy. 2011. *The Precariat: The New Dangerous Class*. London and New York: Bloomsbury.

Thompson, Damian. 1997. *The End of Time: Faith and Fear in the Shadow of the Millennium*. Lebanon: University Press of New England.

Part II

Revised literary genres and visual formats

4 Imagining planetarity in Vandana Singh's speculative short fiction

Klara Machata

The human influence on the planet is a constant presence in today's media landscape and public discourse. Yet, after decades of political agreements, repeatedly missed climate objectives, ever-increasing mountains of scientific evidence and ongoing humanitarian crises in many parts of the world, a sense of disillusionment regarding the future of the planet and of humanity in general is widespread. Media reports, scientific publications, and activist discourse frequently convey a sense of urgency which contrasts starkly with the disorientation and lack of agency that many people feel with regard to their own actions. Timothy Clark (2015) argues that the multiscalar complexity of environmental concerns, such as climate change, confounds dominant notions of individual agency and autonomy (cf. 9–12). Ultimately, this complicates individual and collective action, since environmental matters now seem to entail "thinking of almost everything at once" (78). This level of abstraction is profoundly at odds with the lived experience and the ordinary human scale of perception. Scholars like Annika Arnold (2018) suggest that literature and the arts can help mitigate this disorienting effect by emphasizing the potential of stories to mobilize people and guide their actions (cf. 2). However, whether one can successfully narrate these complex matters and whether these can truly act as a catalyst for change is still up for debate, and it appears that the arts are similarly affected by the deconstructive force of a changing planet. The writer and novelist Amitav Ghosh described this as a "crisis of culture, and thus of the imagination" (9) in 2016. While Ghosh's argument, along with other early publications on climate change in literature, focuses primarily on the novel format,[1] authors of fiction employ various narrative forms and frequently experiment with form and genre in order to engage with the imaginative, ecological, and socio-political challenges of a changing planet.

The US-based Indian writer Vandana Singh is one of many contemporary authors who explore this multitude of planetary entanglements and the changing relationships between humans, nonhumans, and the material

DOI: 10.4324/9781003383598-7

world in their work. While Singh's short fiction is often concerned with climate change, activism, and scientific innovation, it resists didacticism in favor of an approach that highlights the innovative potential and the speculative nature of her writing. Singh herself has an educational background in theoretical physics; she embraces epistemological indeterminacy and explores complex connections across vast temporal and spatial scales. Her work is characterized by formal experimentation, plurality of perspectives, and an ongoing responsiveness to the entanglements of the self and others. As Esterino Adami (2021) observes, the flexibility of Singh's approach and her interest in connecting apparently disparate phenomena allow "a mutual flux between reality and fiction, emerging as a sort of representational mirror that by sketching out dreams and fantasies also grants a scrutiny of life and identity" (17).

In the following, I explore how Singh's short fiction is informed by her interest in the planet, relationality, multiscalarity, and precarity more broadly speaking, before turning to the question of how narrative form can be employed to translate complex systems and entanglements to the reader. Singh's aptly named novella "Entanglement" (2015) serves as an exemplary case study since it showcases many of the concerns and features that are typical of Singh's work. By interweaving five loosely connected plotlines set in different locations all over the globe, the small-scale efforts of individuals trying to fight climate change are projected onto a planetary canvas. The protagonists' success ultimately depends on fragile connections in a chaotic web of relations. Singh employs form to narrativize the interdependences and manifold relationships between humans, nonhumans, and the material world, and urges her readers to look for connections between seemingly independent characters and plotlines. The narrative structure of the text creates an "environmental form" (Morton 2007, 3) with a planetary imagination that turns the readers' attention to various encounters, social relations, and forms of collective action that humans and nonhumans are involved in. Due to this particular form, the story allows a glimpse beyond the limited, human scale of perception and mimics complex spatiotemporal processes that go beyond linear progression. This challenges the reader and their established ways of reading and thinking. The local experiences and individual vulnerabilities of the protagonists merge into a dynamic, multifaceted collage of planetary precarity that highlights the interplay between human and nonhuman, local and global, and social and environmental justice.

From climate change to planetary precarity

The notion of the planetary has recently gained popularity in the environmental humanities in order to address the complex socio-ecological entanglements of our time. Drawing on the work of critics like Édouard Glissant, Masao Myoshi, and Gayatri Chakravorty Spivak, Christian Moraru (2017) defines the planetary as a relational world-system that is characterized by "multiplicity, open-endedness, and sociocultural and political potentialities" (127). Spivak (2003) proposed the planet to overwrite the globe and its totalizing abstraction in order to "imagine ourselves as planetary subjects rather than global agents, planetary creatures rather than global entities" (72–73). This is a useful approach to address the complex socio-environmental entanglements of a rapidly transforming planet, since it rejects universalizing environmentalism and stresses that this planet is a differentiated, political space. While environmental degradation was once largely perceived as a problem that concerned scientists, it is now widely acknowledged to be deeply entangled with neo-liberal and capitalist practices and the reproduction of historical patterns of exploitation that laid the foundation for the economic wealth of the Global North. Imperialism and settler colonialism have destroyed communities and wreaked havoc on their lands, indiscriminately destroying Indigenous knowledge structures and cultures based on cohabitation and deep social-ecological responsibility. This has resulted in multiple, interrelated planetary crises that affect humans, nonhuman species, and the environment both directly and indirectly. In turn, these contribute to increasing inequality so that the social, political, and ethical dimensions of planetary change are undeniable.

Vandana Singh's engagement with the planet in her works is multifaceted: she highlights the connections between humans, nonhuman species, and the planet and shows how these connections exist across vast spatial scales and the social divisions of race, class, caste, gender, age, or species. The vision of the planet that Singh brings forward is social, political, and ecological. In fact, her work is characterized by a broader concern with the exploration of different ways of ethics, knowing and being in the world that resembles Karen Barad's (2007) "*ethico-onto-epistem-ology*" (185, emphasis in original). The attention to complex human-world entanglements and decentering of the subject that Singh frequently engages in are also found in Actor-Network-Theory, Object-Oriented Ontology, and New Materialism. Singh's attention to fragile interdependences aims to find alternative, non-hierarchical ways of relating to one another.

As a productive writer of speculative short fiction, Singh frequently includes other planets, other dimensions, and other-worldly species in her work; yet, rather than employing this otherness purely as a means of distinction, she embraces relationality between humans, nonhumans, and the material world. This, however, does not imply an absence of difference or social division; in fact, her stories frequently focus on marginalized characters and individuals who feel alienated and isolated from others. This can be due to physical isolation, which an imprisoned climate activist experiences in "Arctic Sky" (2022a), but also to ideological differences and estrangement, as in "Entanglement". In several stories, characters note that their own species feels "alien" to them and they seek out animals for companionship. Singh also imagines alternatives in which individuals can escape the constraints that society imposes on them, for example by exploring other dimensions or universes, as in "The Tetrahedron" (2008) and "The Wife" (2008), or by undergoing transformations that culminate in radical otherness. The protagonist in "The Woman Who Thought She Was a Planet" (2008) literally attains planetary form, while in "The Room on the Roof" (2022), a sculptress chooses to let herself become absorbed by the clay of a river bed and becomes one with the material she works with. In "Thirst" (2008), a young mother temporarily transforms into a water-snake and while she eventually returns to her human form, the experience leaves her fundamentally changed. In her acclaimed time travel story "Delhi" (2008), Singh intertwines past and future as the main character Aseem explores the layered histories and potential futures of a megacity. It is the exploration of plurality and connection across vast timescales that is central to this narrative. The speculative nature of her writing allows Singh to capture liminal experiences and to experiment with the borders of reality and fiction, to explore alternative worlds and histories. In doing so, she employs a non-homogenizing, radical openness towards multiplicity and change.

The emphasis on transformation and the imagination of alternatives also applies to those works which are more firmly grounded in reality and overtly concerned with the future of this planet. Despite the central presence of advanced technologies and science in her work, Singh does not resort to the easy techno-fixes that authors of science fiction are sometimes criticized for. The focus of her interest lies in curiosity towards and experimentation with these technologies rather than any supposed "solutions" they could provide and she raises the question of how they can be integrated into a network of different yet connected local responses to planetary change. This is the case in the story "Reunion" (2019), which follows the protagonist Mahua, a formerly working-class Adivasi woman who becomes a scientist. Inspired by the patterns and networks she observes in nature, for example, the veins of a leaf, she develops a sophisticated network of sensors that allows the prediction of extreme weather events and finds a community of underprivileged individuals, who help her to implement her ideas. While Mahua originally struggles with her ancestry and her ageing, brown, female body, she eventually comes to embrace traditional ecological knowledge and makes peace with herself and her role as a planetary creature. By repeatedly returning to the image of the leaf, which appears on the first and final page of the story, Singh weaves a subtle network of association and draws attention to the various structures that organize the planet. In an essay on utopias, Singh attests that she aims "to pay attention to and learn from what is pushed aside, ignored, marginalized" (2022c, 20). This interest in different epistemes and ways of being-in-the-world echoes Spivak's dedication to pluralization and her proposition to embrace "learning to learn from below" as a crucial step "toward imagining planetarity" (2003, 100).

How can we grasp the scientific complexity of human influence on the planet and how can we envision a non-exploitative relationship to the planet? Indigenous cultures and belief systems acknowledged the connections between the well-being of lands and those who populate them long before western scientists began to recognize the destructive impact that the progress-driven idea of modernity had. Modern science has also moved beyond reductionist, mechanistic understandings of the planet and now recognizes that humans are part of the systems they aim to investigate (Barad 2007, 206). In her writing, Singh frequently explores scientific concepts, interrelations, and complex systems,[2] but she also expresses a keen interest in traditional knowledge, grassroots movements, and alternate epistemologies (2022c, 27–34). These interests are represented and frequently enmeshed in her stories. When it comes to communicating scientific matters and complex processes, Matthew Nisbet (2009) argues that "to break through the communication barriers of human nature, partisan identity, and media fragmentation, messages need to be tailored to a specific medium and audience, using carefully researched metaphors, allusions, and examples that trigger a new way of thinking" (15). This leads us to interrogate which formal features these messages exhibit.

Reconsidering form and narrative structure

Many scholars within literary and cultural studies have recently shown a renewed interest in form, which has led to the emergence of a diverse movement called "new formalism".[3] In contrast to the much narrower focus of traditional formalist analyses, in this new understanding the investigation of form encompasses both the formal features of texts and their contextual relations. In his book on forms and story in the Anthropocene, Marco Caracciolo (2021) makes a convincing argument for the investigation of form and the consideration of narrative and natural patterns at this present moment:

> Form provides aesthetic distance from humankind's predicament: through metaphor and imagery; the abstraction of literary form appeases, temporarily, the anxiety of living in times marked by radical, human-induced changes to ecosystems. At the same time, form

> brings out the vast 'fabric' of interconnection in which both humans and nonhumans are intrinsically caught up.
>
> (2)

Form can then be employed to mediate and translate complex matters and to draw attention to connections and interdependencies that usually remain invisible to human perception. It can both guide the reader and startle them, confirm or subvert their expectations, and can make them question previously held assumptions. It is thus clear that if we consider the role of literature and the arts in response to recent planetary changes, questions of form deserve our attention.

While most of the academic discussion on the potential impact of narrative fiction has focused on the novel or on films, short fiction has received comparatively little attention beyond selected close readings of individual texts.[4] It may be counter-intuitive to turn towards a genre that is at least partially defined by its relative brevity when concerned with planetary scales and complex entanglements. Yet, the flexibility, immediacy, and adaptability that are associated with shorter forms of prose fiction also provide unique opportunities to respond to the imaginative and representational challenges that arise on a rapidly transforming planet. As Mary Louise Pratt (1994) notes, short fiction is frequently associated with formal experimentation and used to introduce new, and possibly stigmatized, perspectives and subject matters into the literary arena (cf. 104). Shorter forms are typically more concentrated and frequently rely on suggestion and allusion rather than explicit description, which is why they are considered to be less comprehensive than the novel. Deborah Lilley and Samuel Solnick (2019) argue that, due to their less formally restricted approach, shorter forms are "effectively exempted from some of the responsibility to be exhaustive associated with longer prose forms when they try to address the multiple scales of climate change" (276). This complements the insidious nature of planetary concerns that are "massively distributed in time and space relative to humans", which is why Timothy Morton (2013) famously argues that they "defy overview and resist understanding" (1).

Short fiction is certainly not exempt from all the challenges that this multiscalar interconnectedness and complexity present to storytelling, but rather than relying on realist conventions and plotlines that follow cause-and-effect logic, fragmentation, and non-linearity are much more common and accepted than in traditional long prose fiction. This allows short fiction to "replicate the ways that environmental issues are experienced and registered, and the disjointed, fragmented ways that their effects come to be communicated through different media and discourses" (Lilley and Solnick 2019, 276). In a sense, the resulting narratives are not necessarily *about* environmental issues, such as climate change, but they seem to be written from deep within the socio-ecological entanglements of a planetary age. The fundamentally disorienting complexity of such issues has a strong affective component and is typically associated with feelings of uncertainty and crisis, which some scholars argue the short story is particularly suited to expressing (cf. Virk 2011, 253). This uncertainty in combination with the use of suggestion, implication, and the elimination of intermediary links, all common elements of short fiction, "invite the reader to go on a quest for something that is hinted at, obscured or hidden behind the texture of words" (Patea 2012, 13). Short fiction thus has the potential to encourage active reader participation and demands close attention to what remains left unsaid, to patterns, to ambiguity, and to multiple ways of reading a text.

These features are not *exclusive* to short fiction as many of them can also be found in longer prose narratives, for example, in Richard Powers's *The Overstory* (2018) which has gathered critical attention for its innovative use of form and its rhizomatic structure (cf. Caracciolo 2021, 39–42; Sultzbach 2022, 215–220; Wood 2022, 44–47). Such texts seem to defy categorization just as much as genres overlap and resist universal definition, which is demonstrated by Wood's discussion of *The Overstory* as an example of a short story cycle while most other critics consider it to be a novel. It is worth noting that many features commonly associated with short fiction appear to be particularly suited to the expression of uncertainty, to the representation of marginalized perspectives and non-linear temporality and to drawing attention to fragile interdependences, which are all crucial concerns related to the increasing awareness of a changing planet.

Exploring the entanglement of self and others

As a prolific writer of speculative short fiction, Vandana Singh experiments with form and formal features and employs them in order to explore different ways of knowing and being, to criticize the present, and to articulate alternative futures. While many of Singh's works are informed by her interest in complex systems and relations, the novella "Entanglement" is formally a particularly innovative example of her engagement with the dynamic entanglement of humans, nonhuman species, and the environment. The narrative structure of this comparatively short text mimics a dynamic network that connects seemingly isolated characters in different locations. The individual plotlines explore how climate change affects these different people, their communities, and their environments and what possibilities for individual and collective action they see and engage in. Yet, the true, planetary scale and potential of the narrative are only gradually revealed to the reader as they uncover the fragile, momentary connections between the characters. Fragmentary subheadings precede each section, yet the connection between them and the story is not immediately clear, so that they may confuse rather than guide the reader initially.

The first section focuses on Irene, a scientist of Baffin Island Inuit heritage who works in the Arctic to slow down methane emissions with the help of bacteria and AI technology. Her interactions with a pod of beluga whales spark a journey of self-discovery and remembrance of her heritage and her Inuk name, Enuusiq. In the following section, another researcher, Fernanda, draws inspiration from life-like street art featuring jungle animals in Manaus which helps her to direct public attention to a green roof initiative. In India, a young Dalit boy named Bhola braves an approaching tornado to warn his village and saves other children and a crow in the process. A concerned senior citizen from Texas, Dorothy, joins a group of anti-fracking activists to fight for her grandson's future. In the Himalayas, a student from Shanghai named Yuan searches for a place of learning that appeared in his dreams and finds an abandoned monastery instead, where he saves a monk from an avalanche.

Each of these sections highlights a different aspect of planetary change and when considered in isolation, they appear like a series of snapshots of people in moments of personal or planetary crisis. The final section of the narrative finally confirms what the reader has grown to suspect over the progression of the individual plotlines: that all elements of the story are loosely connected and part of an even grander, macroscalar narrative. Each of the central characters wears a wristlet that runs experimental software. The wristlets gather information about the mood and emotions of their wearers and randomly connect them to another user in moments of high emotional tension. This allows the transmission of brief audio or video messages, but since the software is not fully functional and the connection short-lived, they often surprise the protagonists and can be trivial or helpful. Yet, what remains is a strong emotional effect: when the characters feel desperate or lost, they are reminded that they are not alone. Kelly Sultzbach (2022) points out that the wristlets mimic the notion of quantum entanglement, which Albert Einstein described as a "spooky action at a distance" (quoted in Sultzbach 2022, 222), in which systems can be entangled despite being separated by great distance.

The fragmentary subheadings are similarly entangled with the rest of the text – while they appear relatively nonsensical at first glance, considering them in relation to one another reveals their meaning: they collectively form the sentence "A Butterfly flapping its wings in the Amazon can cause a tornado in Texas" (Singh 2015, 269–322). This refers to the butterfly effect, which Yuan mentions in the story: the idea that a miniscule change in the initial condition of a complex system can trigger chain reactions and thereby lead to a drastically different outcome on a large scale; yet, due to its chaotic nature it is impossible to determine which factors are crucial to a certain outcome. While it is not explicitly stated in the text, what Singh implies here is that this also applies to the small-scale efforts of individuals who engage in climate activism. Due to the frequently imperceptible entanglement across vast temporal and spatial scales, each action contributes to the constant reconfiguration of relations and has the potential to influence the world's unfolding. The final section of the story is introduced by the subheading "*The End*" and further unveils the complexity of the narrative design. It begins in a somewhat paradoxical manner:

> *The End*
>
> The Story Begins
> Or does it end here?
> It ends, the young man thinks, as he climbs the last mountain, emerging into the last alpine valley.
>
> (Singh 2015, 312; emphasis in original)

In this section, the character Yuan, who is initially convinced that he is terminally ill, recovers from his illness with the help of the monk and an unnamed small animal. In return, he saves the monk from an avalanche. Inspired by this experience of connection, he decides to found the place of learning he was searching for in the first place, so that what appeared to be the end turns out to be a new beginning. Due to the narrative design, the reader shares this experience, since this section finally explains the technology of the wristlets. It reveals how the characters are connected to each other and that all subplots are taking place simultaneously. The final two words of the story "... A Butterfly ..." (322), are not only the transcription of a transmission from Fernanda that Yuan's wristlet receives, they also guide the reader back to the first subheading "... Flapping Its Wings ..." (269) in a circular fashion. It becomes clear that the subheadings are arranged in sequence, but since the story begins with the second element and ends on the first, the true beginning is only revealed at the end. The true narrative complexity of the text can only be uncovered if the reader is willing to pay close attention to the hints they receive throughout the narrative. The text invites them to search for parallels and notice loose threads, provided they are willing to let themselves be guided or even misled. If they choose to do so, they themselves become entangled with the text.

While the reader gradually uncovers the macroscalar connections between simultaneous occurrences separated by vast spatial distances, the interplay between entangled entities and dynamic relationships are also central themes within each subsection. This phenomenon is not exclusive to humans: animals and non-sentient beings like bacteria or artificial intelligence apparatuses are involved in these emerging relationships and, as

I have argued elsewhere, their involvement is frequently crucial to how events unfold for the characters. For instance, this can be observed in the interactions between Yuan and the monk's animal companion, and Irene and the whales, respectively (cf. Machata 2022, 54). In addition, Irene's experiments in the Arctic and the complex interconnected systems and self-learning machines she develops and fosters make this particularly clear. They are based on traditional ecological knowledge which she learnt from her grandfather, and the combination of this knowledge, advanced science, self-learning entities, and millions of involved citizen science volunteers leads to a utopian vision of "a multilevel, complexly interconnected grid, a sentience spanning continents and species, a kind of Gaiaweb come alive" (Singh 2015, 278).

The sharp contrast between the scale of the problems the characters face, on the one hand, and their own, limited influence, on the other, is a repeated source of frustration to the characters when considered in isolation. However, as they interact and collaborate with others, they grow to see the world as "an interconnected web of relations – between human and human, human and beast and plant, and all that's living and nonliving" (Singh 2015, 316). This vision is reminiscent of Morton's notion of the "mesh" and its "infinite connections and infinitesimal differences" (2010, 30) and these differences are an important aspect of the story. Despite the central role that connection plays and the generally hopeful tone of the narrative, Singh does not fail to acknowledge the vast inequalities that are amplified by environmental change. Over the course of the narrative, it becomes overwhelmingly clear that the vulnerability of individuals and communities, and the immediacy of their exposure to planetary changes, differ radically. While some characters ponder the fate of future generations in an abstract manner, Bhola and his community have little protection from the approaching tornado and their entire village is destroyed by the storm, which leaves no doubt about the urgency of the planetary crisis. Despite this intense moment of crisis, the Dalit community is not presented as a group of passive subjects: rather, Singh highlights their determination to learn more about climate change and their refusal to sell ancestral lands to a coal-mining company. By contrasting multiple, local, and individual experiences and perspectives and weaving them into a complex, non-linear narrative, Singh creates a multifaceted collage of activism and resistance.

Imagining planetarity

In a discussion of Spivak's planetarity and the act of reading, Hayley G. Toth (2020) stresses the imaginative potential that close reading holds:

> If we as readers observe an ethical sensitivity to the organizing structures of the text, we may participate in a denaturalization of our existing

> ways of being and knowing, and be motivated to alter – conceptually and practically – the ways in which we world-make outside of the act of reading.
>
> (462)

Thus, reading has the potential to become a highly political act which involves not only a critical gaze onto the status quo, but the exploration of potential alternatives – it encourages the reader to participate in the co-creation of the world's unfolding. It is safe to assume that not every reader will be equally receptive to this deep level of engagement with the texts they encounter, and the act of reading alone will change little for those most heavily affected by planetary environmental change. Yet, if we return to considering the planetary environmental crisis as one that entails a cultural and imaginative dimension, reading as suggested above, can foster critical reflection and allow new perspectives on the past, the present, and potential futures and thereby contribute to a broader paradigm shift.

Vandana Singh's speculative short fiction frequently discusses key features of complex systems and addresses the disorienting and alienating effects that an awareness of this complexity can have, particularly in face of multiple planetary crises. Her work also highlights the opportunities for social and political change which arise out of the entanglement and intra-action between humans and the nonhuman world. She explores alternatives to linear progression and cause-and-effect logic not only thematically but also aesthetically, owing to her innovative use of narrative form. Singh uses the short form and its potential to capture fleeting and fragile experiences of connection masterfully and employs a non-totalizing gaze that exposes the manifold effects and the hybridity of affected beings and communities. She draws attention to marginalized individuals and perspectives and to multiple ways of recognizing the animacy and value of the planet, all of which are still frequently neglected in mainstream western environmental discourse.

The characters in "Entanglement" become aware of their entangled histories and of the consequences of their actions as they interact with other beings. At the same time, they also draw inspiration from Indigenous teachings, Hindu spirituality, art, science, education, family ties, community, and artificial intelligence. While the advanced technology developed by Yuan provides the infrastructure for the audible transmissions between the characters and allows them to connect to other people across vast spaces and differences in age, class, language, and nationality, the fragmentary short-time transmissions can be confusing to both the characters in the story and to the reader. These fleeting links are usually triggered in moments of desperation or crisis and their strong affective component can jolt receiving characters into action. However, the sending characters typically remain unaware of their reactions and of the further unfolding of events. It remains up to the reader to make sense of how the individual plotlines and transmissions relate to one another and to uncover the fragile connections in a dynamic and unstable web of constant refiguration. The particular perspective of the reader allows intimate insight into the motivations and struggles of individuals and grants aesthetic distance that facilitates the recognition of similarities, differences, patterns, and forms that organize a text.

As Adami attests, one of the most striking features of Singh's stories is that they "involve the reader in projecting meaning and participating in larger processes of comprehension, imagination and representation of the complexity of the postcolonial world" (2017, 99). By using form and formal features to underline the relationality and multiscalarity central to the complex socio-ecological entanglements, Singh invites the reader to pay close attention to texts and their organizing principles. Although not every reader will follow the suggested paths or recognize the underlying ideas and structure, the multiple layers and perspectives in Singh's work make a strong point for collaboration, empathy, and the intersection of environmental concerns and social justice. The resulting ethical implications urge every reader to question their own position in, as well as their own assumptions about, the planetary network of entanglement. Reading stories like Singh's challenges the reader to engage in a process of critical reflection and to actively participate in the co-creation of new understandings, which empower individuals and inspire a sense of responsibility. Ultimately, these emerge as crucial steps towards imagining planetarity and planetary solidarity.

Notes

1 See also Adam Trexler (2015), Antonia Mehnert (2016), and Adeline Johns-Putra (2019).
2 Complex systems are biological, physical or social systems with many interacting components. These systems often possess feedback loops, tipping points and other non-linear behavior which occurs over multiple scales. Consequently, patterns and large-scale behaviors cannot be determined by examining isolated parts of the system so that the focus of investigation lies on the relations themselves rather than on individual components (cf. Siegenfeld and Bar-Yam 2020, 12–13).
3 See, for example, Marjorie Levinson (2007) or Elizabeth Kovach, Imke Polland and Ansgar Nünning (2021) for a collection of different approaches.
4 Notable exceptions are Deborah Lilley and Samuel Solnick's chapter (2019, 272–290), and Marco Caracciolo (2021) which, while not focused on short fiction exclusively, provide an in-depth discussion of form and story in the Anthropocene. Niamh Wood (2022) advocates for the short story cycle's capacity to defamiliarize climate change (44–47) even though the examined text is most commonly classified as a novel.

References

Adami, Esterino. 2017. "The Shapes of Things to Come in Indian Science Fiction. Narrative and Style in Manjula Padmanabhan and Vandana Singh." In *Other Worlds and the Narrative Construction of Otherness*, edited by Esterino Adami, Francesca Bellino, and Alessandro Mengozzi, 85–100. Milan: Mimesis International.

Adami, Esterino. 2021. "Pushing the Boundaries of Genre and Gender: A Stylistic Investigation of Vandana Singh's Science Fiction Stories." In *Intersezionalità e genere* [Intersectionality and gender], edited by Anna Mongibello and Katherine E. Russo, 15–34. Trento: Tangram Edizioni Scientifiche.

Arnold, Annika. 2018. *Climate Change and Storytelling: Narratives and Cultural Meaning in Environmental Communication*. London: Palgrave Macmillan.

Barad, Karen. 2007. *Meeting the Universe Halfway. Quantum Physics and the Entanglement of Matter and Meaning*. Durham, NC: Duke University Press.

Caracciolo, Marco. 2021. *Narrating the Mesh: Form and Story in the Anthropocene*. Charlottesville, VA: University of Virginia Press.

Clark, Timothy. 2015. *Ecocriticism on the Edge: The Anthropocene as a Threshold Concept*. London: Bloomsbury Academic.

Ghosh, Amitav. 2016. *The Great Derangement: Climate Change and the Unthinkable*. Chicago, IL: University of Chicago Press.

Johns-Putra, Adeline. 2019. *Climate Change and the Contemporary Novel*. Cambridge: Cambridge University Press.

Kovach, Elizabeth, Imke Polland, and Ansgar Nünning, eds. 2021. *Forms at Work*. Trier: Wissenschaftlicher Verlag Trier.

Levinson, Marjorie. 2007. "What Is New Formalism?" *PMLA* 122 (2): 558–569.

Lilley, Deborah, and Samuel Solnick. 2019. "The Short Story and the Environment." In *The Edinburgh Companion to the Short Story in English*, edited by Paul Delaney and Adrian Hunter, 272–290. Edinburgh: Edinburgh University Press.

Machata, Klara. 2022. "Entangled Climate Heroes: Heroic Agency and Intra-action in Vandana Singh's 'Entanglement'." In "Climate Heroism", edited by Stefanie Lethbridge and Anna Hemkendreis, special issue of *helden. heroes. héros. E-Journal zu Kulturen des Heroischen* 8: 49–56. https://doi.org/10.6094/helden.heroes.heros./2022/CH/06

Mehnert, Antonia. 2016. *Climate Change Fictions: Representations of Global Warming in American Literature*. London: Palgrave Macmillan.

Moraru, Christian. 2017. "'World,' 'Globe,' 'Planet': Comparative Literature, Planetary Studies, and Cultural Debt after the Global Turn." In *Futures of Comparative Literature*, edited by Ursula K. Heise, 124–133. New York City: Routledge.

Morton, Timothy. 2007. *Ecology Without Nature: Rethinking Environmental Aesthetics*. Cambridge, MA: Harvard University Press.

Morton, Timothy 2010. *The Ecological Thought*. Cambridge, MA: Harvard University Press.

Morton, Timothy. 2013. *Hyperobjects: Philosophy and Ecology after the End of the World*. Minneapolis, MN: University of Minnesota Press.

Nisbet, Matthew C. 2009. "Communicating Climate Change: Why Frames Matter for Public Engagement." *Environment* 51 (2): 12–23.
Patea, Viorica. 2012. "The Short Story: An Overview of the History and Evolution of the Genre." In *Short Story Theories: A Twenty-First-Century Perspective*, edited by Viorica Patea, 1–24. Leiden: Brill.
Powers, Richard. 2018. *The Overstory*. New York and London: W.W. Norton & Company.
Pratt, Mary Louise. 1994. "The Short Story: The Long and the Short of It." In *The New Short Story Theories*, edited by Charles May, 91–113. Athens, OH: Ohio University Press.
Siegenfeld, Alexander F. and Yaneer Bar-Yam. 2020. "An Introduction to Complex Systems Science and its Applications." *Complexity* 2020: 1–16.
Singh, Vandana. 2008. "Delhi." In *The Woman Who Thought She Was a Planet: And Other Stories,* 19–38. Delhi: Zubaan and Penguin Books India.
Singh, Vandana. 2008. "The Tetrahedon." In *The Woman Who Thought She Was a Planet: And Other Stories,* 139–167. Delhi: Zubaan and Penguin Books India.
Singh, Vandana. 2008. "The Wife." In *The Woman Who Thought She Was a Planet: And Other Stories,* 168–179. Delhi: Zubaan and Penguin Books India.
Singh, Vandana. 2008. "The Woman Who Thought She Was a Planet." In *The Woman Who Thought She Was a Planet: And Other Stories*, 39–54. Delhi: Zubaan and Penguin Books India.
Singh, Vandana. 2008. "Thirst." In *The Woman Who Thought She Was a Planet: And Other Stories,* 89–108. Delhi: Zubaan and Penguin Books India.
Singh, Vandana. 2015. "Entanglement." In *Loosed Upon the World: The Saga Anthology of Climate Fiction*, edited by John Joseph Adams, 269–322. New York City: Saga Press.
Singh, Vandana. 2019. "Reunion." In *The Gollancz Book of South Asian Science Fiction*, edited by Tarun K. Saint, 341–365. Gurgaon: Hachette India.
Singh, Vandana. 2022a. "Arctic Sky." In *Utopias of the Third Kind*, 9–18. Oakland, CA: PM Press.
Singh, Vandana. 2022b. "The Room on the Roof." In *Utopias of the Third Kind,* 71–96. Oakland, CA: PM Press.
Singh, Vandana. 2022c. "Utopias of the Third Kind." In *Utopias of the Third Kind*, 19–36. Oakland, CA: PM Press.
Spivak, Gayatri Chakravorty. 2003. *Death of a Discipline*. New York: Columbia University Press.
Sultzbach, Kelly. 2022. "More-than-Human Collectives in Richard Powers' *The Overstory* and Vandana Singh's 'Entanglement'." In *The Cambridge Companion to Literature and Climate*, edited by Adeline Johns-Putra and Kelly Sultzbach, 214–226. Cambridge: Cambridge University Press.
Toth, Hayley G. 2020. "Spivak's Planetarity and the Limits of Professional Reading." *Comparative Critical Studies* 17 (3): 459–478.
Trexler, Adam. 2015. *Anthropocene Fictions: The Novel in a Time of Climate Change*. Charlottesville, VA: University of Virginia Press.
Virk, Tomo. 2011. "The Short Story as a Genre of Ontological Uncertainty." *Forum for World Literature Studies* 3: 248–254.
Wood, Niamh. 2022. "Realism in Eco Fiction: Climate Change and the Short Story Cycle." *Social Alternatives* 41 (3): 43–47.

5 Precariousness and resistance

Petro-despotism and the imaginative power of literature

Chiara Lanza

Introduction: the elephant in the room

Our lives are saturated with oil: not only is it involved in the movement of goods and people, but oil has become necessary for producing the objects we use and interface with daily, from our phones to the t-shirts we wear. If, until some decades ago, it was easy to forget about the ubiquitous presence of fossil fuels, today it has become impossible not to mention them when thinking about the precariousness of our lives on Earth: the connection between emissions from carbon majors and temperature increase cannot be overlooked. For these reasons Marco Grasso, professor of International Environmental Policy at the University of Milan-Bicocca, and the journalist Stefano Vergine, call international and national oil companies the big "elephant in the room" (2020, 244): in spite of their causal and moral responsibility, oil companies are virtually absent from international conferences on climate change, and no strong action is taken to prevent them from continuing, or even increasing, their activities of exploration and extraction, masked under the guise of an ecological transition.

In such a controversial context, the arts and literature have an important role to play. The disruptive actions of activist groups like Just Stop Oil, ranging from protesters tying themselves to goalposts (Pandey 2022), to activists gluing their hands to reproductions of Da Vinci's paintings (Gayle 2022), are provocative, but their underlying meaning is not always fully grasped. News reports often fail to convey the purport of these initiatives, thus participating in "concealment cultures" that, in collusion with the carbon economy, blur people's imaginations and "damage the health of the planet" (Dwivedi 2023, 5). Perhaps, were the actions of environmental campaigners approached as artistic performances rather than simple protests, they would stir that "ontological appeal to intervene", that, according to Santiago Zabala (2019, 21), is at the core of contemporary art. In other words, the relation with the work of art in our age is not only aesthetic but also active and practical, and able to stimulate questions about

DOI: 10.4324/9781003383598-8

the meaning of existing and acting in a world in crisis. Hence, the spaces that artistic and literary narratives open up can help us live, interpret, and understand in ways that differ from those imposed by the mono-cultures of concealment that surround us.

Literature operates from a privileged position, because novels and poems are, for the most part, structured in such a way as to make meanings and intentions not only intelligible but relatable. As Pieter Vermeulen (2020) writes in *Literature and the Anthropocene*, "literature's affective, narrative, imaginative, and self-reflexive powers make it a vital site for coming to terms with the challenges of the current environmental crisis" (29). Moreover, not only can readers come to terms with current challenges, but the world-building capability of narratives makes it possible to imagine different pasts or envisage alternative future trajectories to present experiences of precariousness and resistance. Petrofiction is a case in point: while it recounts the socio-environmental precariousness of lives affected by oil extraction, it also offers spaces to grasp the consequences of action or inaction, resistance or compliance, and in which to build alternative scenarios of hope.

The present chapter aims at discussing some of the "cultures of concealment" through which the carbon economy has operated so far, and to put them in perspective through one of the most recent publications in the field of petrofiction, namely, Imbolo Mbue's (2021) *How Beautiful We Were*. Short-listed as a finalist for the 2022 PEN/Faulkner Award for fiction, this novel has reached a wide audience. Unwrapping some of its core themes and messages is useful for understanding how fiction provides an additional layer to past and current events, and how its portrayals of precarious lives and attempts to resist are relevant to ongoing battles for equity and safety. Consequently, concepts such as those of precariousness, resistance, and hope will be drawn upon as a lens through which to compare and contrast Mbue's novel with events that have marked movements for socio-environmental justice. The resulting analysis aims to provide insights into the key role played in the novel by rootedness to place and familial ties in dealing with the challenges and injustices faced by communities affected by oil extraction.

The responsibilities of Big Oil

Climate change and environmental justice intersect on many levels. In particular, with reference to fossil fuels, their entanglement calls for a closer look at the contribution and responsibility of Big Oil. Numerous studies have been published on the question of carbon majors' emissions. For instance, the 2017 Carbon Majors Report shows how, since 1988, just 25 corporate and state-producing entities have been responsible "for 51%

of global industrial GHG emissions", with the percentage increasing to 71 percent when 100 producers are taken into account (Griffin 2017, 8). Fossil fuel producers and investors might indeed "hold the key to tackling climate change" (Riley 2017), but if they are not seriously involved in any debate and effective policy, then ecological goals cannot be achieved. Similarly, the Climate Accountability Institute, established in 2011 and engaged in "research and education on anthropogenic climate change, [...] and the contribution of fossil fuel producers' carbon production to atmospheric carbon dioxide content" (Climate Accountability Institute 2022), published a Chart of Carbon Majors and Global Emissions from 1810 to 2017, in which the way that the curve of emissions from carbon majors approximates and increases at the same pace as the one of total emissions from fossil fuels and cement can be observed (see Figure 5.1).

Despite all the evidence that the trend has to be reversed in order to fight climate change, there is still debate. The outcomes of the 26th Conference of the Parties (COP26) in November 2021 constitute a case in point. What seemed a successful accomplishment in the draft agreement, which called upon the Parties for the "phaseout of unabated coal power and of inefficient subsidies for fossil fuels" (CNN 2021), turned into a disappointing and nonbinding "phasedown of unabated coal power and phase-out of inefficient fossil fuel subsidies" (UNFCCC 2021). Thus, the pill was easier

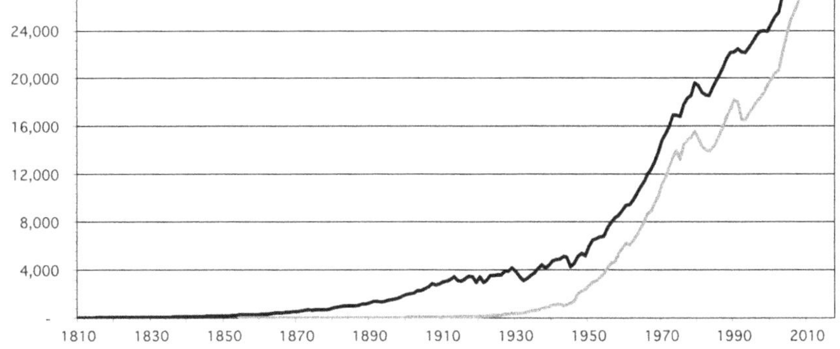

Figure 5.1 Chart of Carbon Majors and Global Emissions, 1810–2017 (Climate Accountability Institute 2020).

to swallow for all the Parties that wished to continue their investments in fossil fuels.

Apart from being causally responsible for greenhouse gas emissions, oil companies have a moral obligation too. This stems from the fact that the effects of extracting and burning fossil fuels have been known since at least 1990, when the first assessment report by the Intergovernmental Panel on Climate Change (IPCC) came out. As stated in the policymaker summary of working group III (Formulation of Response Strategies), the consumption and production of "energy from fossil fuels (coal, petroleum and natural gas excluding fuel wood) [...] accounts for an estimated 46%" of total sources of anthropogenic greenhouse gases (IPCC 1992, 115). However, the 1990 report was the tip of the iceberg: research in this field had been a work-in-progress for a couple of decades already, sometimes funded by oil companies themselves (Grasso and Vergine 2020, 448). Scientists of the Oak Ridge National Laboratory, a research center directly financed by the US Department of Energy, claim in their introduction to a document published in 1982 that "it may become necessary to try to limit the concentration of atmospheric CO_2", and, linking temperature rise to increased CO_2, warrant an "intensive development of non-fossil energy sources" (Perry et al. 1982, 991). Despite such reports, oil and coal companies have not stopped lobbying to keep their interests intact, even thriving, as highlighted by the latest assessment report by the IPCC. Their ability to "block initiatives towards decarbonisation" and to "shape government action on climate change" represents a real constraint to mitigation efforts (Grubb et al. 2022, 28).

In a 2020 interview for the *Time* magazine, Shell CEO, Ben van Beurden, was asked about "Shell knew", the activist slogan used to accuse the company "of failing to act on climate change despite knowing the consequences", to which he responded: "Yeah, we knew. Everybody knew. [...] And somehow we all ignored it" (quoted in Worland 2020). What is striking in such a declaration is not only the tone, but especially the language that smacks of domineering ego, ignoring the pressing global concern. Instead of answering "Yeah, Shell knew", van Beurden uses a noncommittal first-person plural, reinforced by "everyone" and "we all" in the second part of the statement, thus including the audience too. By doing so, he reiterates a long-standing argument, according to which consumers too are responsible for global emissions, and the company only answered a pre-existing demand. As Grasso and Vergine explain in *Tutte le Colpe dei Petrolieri* (All the Fault of Oil Companies), this reasoning is fallacious for a number of reasons: on the one hand, the demand increased because alternatives to fossil fuels became less viable and more expensive; on the other hand, it is not easy for consumers and stakeholders to make responsible choices when information is withheld or, simply not accessible, as was the case before the first IPCC report was published (2020, 159–168). In other words, before 1990 it was not common knowledge that the activities of Shell, as well as of other carbon majors, were contributing to global warming, but Shell did know and did nothing to inform either stakeholders or consumers. Moreover, daily field production of oil has risen constantly, from around 8 million barrels per day in the US alone at the beginning of the 1980s to almost 13 million in 2020, prior to the pandemic (US Energy Information Administration).

However, the environmental damage caused by the activities of oil companies is not limited to greenhouse gases emissions. Exploration, extraction, and production are known to expose extraction sites, both offshore and onshore, to great risk, as they can have catastrophic consequences on both the environment and the well-being of people living in the affected areas. For instance, the images of the Deepwater Horizon disaster in 2010 are still very vivid in many people's minds, with its terrible repercussions on marine life. The exposure to risk becomes particularly problematic in poorer parts of the world, where international oil companies such as Shell and Chevron invested throughout the second half of the 20th century, as their proceedings were less likely to be closely monitored. Outdated equipment was installed, subsequently leading to explosions and frequent oil spills, which in turn caused the destruction of "all wildlife, and plant life", poisoned "the atmosphere and therefore the inhabitants" (Saro-Wiwa 1992, 80–81). These words can be read in Ken Saro-Wiwa's 1992 book titled *Genocide in Nigeria. The Ogoni Tragedy*, an account of the slow violence done to a land and its dwellers, that is, a violence that is "incremental and accretive", with effects spreading over an extended period of time (Nixon 2011, 2). The devastation caused by the activities of oil companies in the Niger Delta region goes beyond reckoning, reminding us of the "resource curse": that a "favourable natural resource endowment" does not necessarily correspond to wealth and benefits for a country. In fact, especially in authoritarian contexts, the abundant presence of a resource could turn into a curse, thus alienating people from their land (Auty 1993, 1).

The power of literature: Mbue's *How Beautiful We Were*

In 1992, the same year in which *Genocide in Nigeria* was published, Amitav Ghosh published in *The New Republic* an article titled "Petrofiction: the oil encounter and the novel". In this article, Ghosh compares the Oil Encounter to the Spice Trade for its historical resonance, but meditates on the possible reasons for its imaginative sterility: if the latter nurtured a vast literature, the same could not be said about the former. "The truth", the author claims, "is that we do not yet possess the form that can give the Oil Encounter a literary expression" (1992, 31), because the reality of oil is baffling, slippery, and, in the 1990s, it was already starting to sound controversial. After such premises, Ghosh proceeds to review a notable exception to his argument, that is, Abdulrahman Munif's (1984) *Cities of Salt*, first published in 1984 and later translated into English by Peter Theroux. Since 1992, however, much has changed, and literature on the subject has burgeoned: "writers, artists, and filmmakers have taken up the challenge of producing the petrofictions we need to fully make sense of oil societies", wrote Canadian scholar Imre Szeman (2012, 3). Scholarly articles and anthologies on the subject have also been published, testifying to the ubiquity of oil in both our societies and on the literary scene. For instance, *Oil Fictions: World Literature and Our Contemporary Petrosphere* (2021), edited by Stacey Balkan and Swaralipi Nandi, offers a variety of articles and essays on the social and cultural pervasiveness of oil with a specific focus on countries of the Global South. Imre Szeman himself has contributed greatly to giving shape to the vast corpus on fossil fuels; in 2017, he co-edited *Energy Humanities*, an anthology aimed at addressing "the social as well as environmental consequences of energy once it gets industrialized across the globe" (Szeman and Boyer 2017, 1), in the belief that the humanities hold the key to understanding and interpreting such consequences and the much-needed intervention.

One of the most recent novels written on the topic is *How Beautiful We Were* (2021) by the Cameroonian author Imbolo Mbue. The story is set in a fictional African village called Kosawa, and follows the villagers' battle against the American oil company Pexton from the 1980s to 2020. The novel opens with a village gathering in the main square, dominated by an ancient mango tree, with three representatives of Pexton, who routinely tour all the villages in the area to answer questions and praise the achievements of the company they work for. None of them speaks the local language, and the village chief Woja Beki acts as interpreter and mediator; however, no one trusts the chief's words after Pexton gifts him with a new, luxurious house, and provides his sons with government jobs. With villagers gathered to listen to three white men celebrate the benefits of oil extraction, this episode brings to mind other occurrences in the long history of exploitation of poorer countries. One such example is the speech of the US Navy Commodore Ben Wyatt in 1946, who travelled to the Marshall Islands in the Pacific Ocean to ask for the inhabitants' permission to use their atoll for nuclear weapon testing. He justified his request with the claim that they would be acting for "the good of mankind" (Brown 2014, 40). However, Wyatt did not actually need the permission of local inhabitants: the Bikini Atoll had already been approved by President Truman as a nuclear testing site (40–44). A few decades later and in a very different area of the world, similar dynamics are imaginatively re-enacted in Mbue's novel: promises of wealth and development are used to mask a stark reality of exploitation, environmental degradation, and resource depletion. The Pexton men show no intention of engaging in a productive dialogue and deny any implication in the plight affecting Kosawa. In fact, Pexton started its operations of extraction years before the village gathering which opens *How Beautiful We Were*, and the consequences of gas flares and frequent oil spills have become evident:

> I think about how the air and water of Kosawa progressed from dirty to deadly [...] whatever life was left in the big river disappeared. Within a year, fishermen broke down their canoes and found new uses for the wood. Children began to forget the taste of fish. The smell of Kosawa became the smell of crude. The noise from the oil field multiplied; day and night we heard it in our bedrooms, in our classroom, in the forest. Our air turned heavy.
>
> (Mbue 2021, 32)

There are many similarities between the events narrated by Imbolo Mbue and the accounts of the devastation of the Niger Delta and its inhabitants which can be read in Ken Saro-Wiwa's essays and books. In *Genocide in Nigeria*, already mentioned, Saro-Wiwa reproduces a petition written in April 1970 by Ogoni leaders and addressed to the military governor of Rivers State which claims that "large acres of mangrove swamps [...] have been destroyed by the periodic out-flow of crude oil into our rivers and streams, which have killed [...] fishes and crabs, mudskippers, oysters, shell-fishes" and that "most inland waters, rivers and water courses have today been polluted by crude oil". Not only was clean water not accessible anymore, but "large stretches of rich agricultural land" were grabbed or destroyed by Shell-BP too (1992, 45–46). The inhabitability and precariousness of such an environment contrast deeply with the former lifestyles and traditions of the villagers described in the novel, who take pride in their origins and fertile land, and think of themselves as indomitable "sons [and] daughters of the leopard" (Mbue 2021, 30).

With slim hope left for better atmospheric conditions due to the ongoing extraction undertaken by Pexton, Kosawans at first try to adapt and eke out a living by hunting in the forest. Most of the ingredients that have become scarce, because of "dying farms and emptying barns" (Mbue 2021, 53), are bought at a market in the nearest city, which implies paying inflated prices for primary commodities and, for the women, making a long bus ride back and forth. Very soon, polluted water and land and unbreathable air start affecting physical health too. The children seem particularly afflicted, as they are the first to develop coughs and high fevers. Death and fear spread: "We feared someone in our homes would catch this sickness that had arrived like a thief in the dark and was now hovering outside every hut, waiting for its chance to enter" (Mbue 2021, 8). There is indeed a causal link between oil spills and infant mortality. Anna Bruederle and Roland Hodler (2019) conducted a study on the health hazard of onshore oil spills in Nigeria; in their article, they write that:

> When crude oil or other petroleum products leak into the environment, the different compounds (depending on their physical properties) evaporate into the air, are absorbed by the soil, or enter ground and surface water. [...] Hazards to human health may result from dermal contact with soil and water; ingestion of contaminated drinking water, crops, or fish; or inhalation of vaporized product or PM and partly burned hydrocarbons produced by fires.
>
> (5467)

In addition, Bruederle and Hodler found evidence that "oil spills likely not only affect neonatal and infant mortality, but also increase the risk of abortion and stillbirth", and they conclude that "the frequent onshore oil spills in Nigeria are the cause of an alarming ongoing human tragedy" (2019, 5470). Hence, the events related by Mbue might be fictional, but they are grounded on real happenings. As a consequence, readers are made aware of ecological and human catastrophes that have taken or are taking place in other regions of the world through a narrative intervention, which is capable of stirring the affective and imaginative responses needed to understand and face the challenges of our precarious age of crisis.

The connections between Kosawa and real villages in the south-eastern area of Nigeria are not limited to the calamitous consequences of oil spills for both the environment and human health. For example, the novel's protagonist, Thula Nangi, seems to embody some characteristics of leading African figures in their attempts to resist and fight socio-environmental injustice, within and outside Nigeria. At the outset Thula is part of "The Children", a group of peers who offer a multifocal perspective on the main events that dot the lives of villagers. The novel interestingly alternates chapters with a conventional structure (with a first-person singular narrator) with chapters in which a communal "we" conveys the collective voice of a whole generation, grown up amidst gas flares and oil spills. Thula's voice, however, stands out from the chorus and qualifies itself as more reflective, and deeply committed to seeking truth and justice. She is indeed affected from a young age due to the active engagement of her father and his disappearance in the city of Bézam, to which he travelled to ask the local government for help against pollution and waste dumping by the foreign oil company. One episode, called "the massacre", changes her life: after three Pexton representatives are kidnapped, soldiers arrive in Kosawa and shoot at the crowd of men, women, and children without discrimination. An American news reporter and photographer, Austin, manages to take some pictures and send them to foreign newspapers, thus exposing the state-corporate nexus, and undermining the reputation of Pexton. International mobilization results, some years later, in the opportunity for Thula to pursue her college education in the US, thanks to a scholarship instituted to help local children. Her years spent abroad change her profoundly and turn her into the leader that Kosawa needed: she exchanges letters with the other children who, having grown up, now understand the need to give a real shape and direction to their resistance. Once back in her native country, Thula is determined to seek justice and lead her people to freedom from Pexton and an exploitative government. A parallel can be drawn between Thula and Wangari Maathai, co-founder of Kenya's Green Belt Movement and the first African woman to win the Nobel Peace Prize. Like Maathai, Thula too studies abroad thanks to a scholarship, learns about civil disobedience, joins civil rights campaigns, and finally decides to return to her own country to fight environmental injustice and the consequent erosion of civil rights.

In *Slow Violence and the Environmentalism of the Poor*, Rob Nixon (2011) devotes some pages to Maathai and the intersectional nature of the Green Belt Movement, which, while mobilizing against the slow violence of deforestation and soil erosion, also asked for security, territorial integrity, and rights in the context of a militaristic and misogynistic regime (128–149). Even though *How Beautiful We Were* is set in a fictional country whose political context is never directly commented on, it is possible to imagine parallels between Kenya's government and the authoritarian rule of Thula's country, where corruption and state-sponsored violence seem widespread. Thula, however, is not considered a real threat by the government and continues teaching her students about resistance, democracy, and environmental justice: "The government yawned when they heard about her Village Meeting in her government-provided house. They said: what can one angry woman do?" (Mbue 2021, 342). Maathai, Nixon writes, vilified by the Moi regime, was similarly "represented as overly emotional and unhinged, an unnatural woman, uncontrollable, unattached, without a husband to rein her in and keep her (and her ideas) respectable" (2011, 146). Thula's story is fictional, but it resonates with truth: the need to label women as either too emotional or as dangerously overqualified seems to be a common thread connecting different figures across time and space. Moreover, the fact that she does not have children raises doubts about her ability as a leader, to the point that her closest friends, convinced that "nothing could make her respectable besides motherhood and marriage" (Mbue 2021, 307), drug her and stage a night ceremony to propitiate the spirits, hoping a (male) child could grow in her to take on the battle in her stead.

Thus, Thula's life is illustrative of how "patriarchy and petro-capitalism have worked in tandem" (Deckard 2021, 44) in petro-states through the regulation and control of the bodies of women, whose motherly and wifely role is deemed essential in sustaining and feeding petro-masculinity. Thula's friends prove unable to see these links, and end up reiterating the very dynamics and mindsets that make petro-imperialism possible.

A further and visible connection between the protagonist of *How Beautiful We Were* and historical figures in the fight for justice is found in that between Thula and Ken Saro-Wiwa. In particular, the latter's death might be a key to interpreting the ending of the novel. As Ike Okonta and Oronto Douglas (2001) write:

> Saro-Wiwa recognized the adverse ecological impact of Shell's operations and began pseudonymously to write protest letters. [...] In 1992, now spokesman of the Movement for the Survival of the Ogoni People (MOSOP), Saro-Wiwa attacked [Shell]'s gas-flaring practices directly at the Tenth Session of the Working Group on Indigenous Populations in Geneva on July 28.
>
> (78)

Saro-Wiwa's activities as ambassador of the Movement for the Survival of the Ogoni People resulted in his detention in the Port Harcourt Prison in 1993; after his release, he was arrested again in 1994, accused of "incitement to murder" and hanged on November 10, 1995, together with eight other MOSOP members (Boyd 1995, vii–xiv). Circumstances between fiction and fact differ greatly, as Thula in the story is neither arrested nor hanged; however, her position as proponent of a movement called "Liberation Day" and her death with close friends and collaborators are reminiscent of the events that marked the last years of Saro-Wiwa's life, especially in the way their execution takes place, without any witnesses or official trial. *How Beautiful We Were* ends with Thula's one-year death celebration and some closing remarks on the last days of Kosawa and the scattering of its former inhabitants. Pexton digs "a new well in what used to be [the] village square" (Mbue 2021, 357), uprooting the mango tree, a symbol of union and peace, and a material entity around which the village's life revolved. No mention is made of possible positive developments. In spite of that, the connection between Thula and Ken Saro-Wiwa might leave some room for hope. Ten years after Saro-Wiwa's death, "conditions across the oilfields remained the same, only worse" (Watts 2008, 37); yet, his son's perseverance also led to some victories:

> He took up the cause as writer and speaker on the international human rights circuit but also, critically, as a lead plaintiff in a fourteen-year-long

> case against Shell for complicity in his father's execution and for paying soldiers who had committed human rights abuses in Ogoniland. On June 9, 2009, days before the trial was due to begin in New York, Shell settled out of court, agreeing to pay $15.5 million, mostly into a trust fund for the Ogoni people.
>
> (Nixon 2011, 125)

Shell's decision to settle out of court and create a fund for the Ogoni people was indeed a moral victory, but not a legal one. By settling the case before it began the company did not provide any legal precedent, and was not legally bound to operate any clean-up or to stop extracting in the Niger Delta (125–126). Despite that, this modest outcome was a success for Saro-Wiwa Jr., and a turning point in the pursuit of environmental justice and human rights. By reading Mbue's novel against these events, it is therefore possible to envisage a similar follow-up to the story: even if at the end of the novel Kosawa is abandoned and seemingly forgotten by the new generation of children, hope is not unconceivable. Legal battles for environmental justice are ongoing, and, because of that, there is space for a story about a future Kosawa, which its children have cleaned and rebuilt, and where precariousness has turned into stability and safety. The effects of resistance might be delayed but, as long as people are willing to fight and to tell stories about their fights, regeneration is still possible.

Conclusion: possibilities of hope

How Beautiful We Were is open-ended and, because of that, full of fruitful possibilities. Throughout the novel, a strong sense of community, the importance of family ties, and the ancestors' teachings are elements that are continuously stressed, and leave readers in the post-2020 era and in distant countries with a feeling that precariousness can be fought when rootedness and collaboration are present. The reasons why protagonists of the story fail to preserve their lands and community are mostly a lack of education and several poor choices made by Thula's friends. Mbue herself, in an interview with David Ebershoff, stated that her characters show a unique ability to hope and a desperation to save their land (The Center for Fiction 2021), assertions that could hold the key to interpreting the novel and its ending as inspiring rather than dark and dismaying. The author is not offering ready-made or comfortable solutions either, as the path followed by Thula is long and full of insurmountable obstacles: she files lawsuits to a US court against Pexton based on the Alien Tort Statute – as Saro-Wiwa Jr. did successfully – that are dismissed, and her Liberation Day movement does not stir much of a reaction by the local government. However, not all options are explored, such as appealing to European courts or involving NGOs and global justice movements in order to mobilize international support, options that are still open at the end of the novel. As Rebecca Solnit (2016) states in *Hope in the Dark*, hope constitutes a "grounds to act", because it provides "the foundation for an ongoing series of acts of defiance, those acts necessary to bring about some of what we hope for while we live by principle in the meantime" (110). In this sense one might claim that Mbue's novel could be read as a road map to facing future challenges and to making our Earth more sustainable: by making readers aware of the obstacles, it also empowers them to imagine alternative narratives of collaboration and of ethical commitment to achieve common goals, such as justice, peace, and a safe and healthy environment to dwell in and coexist with more-than-humans.

Thus, literature emerges from such an analysis as an indispensable guide in our era of precarious dwelling and being on the planet. Novels such as Mbue's, as well as other forms of art, can contribute to giving a de-centered and dramatized perspective on phenomena otherwise hard to interpret or comprehend, either because of their scale or for their remoteness in time and space. Graphs and articles, such as those mentioned at the beginning of the present chapter, record and quantify; the arts are needed to narrate, describe, and to imagine possibilities of action. As Adam Trexler (2015) claims in *Anthropocene Fictions*,

> climate science can provide a high level of certainty about the dangers of burning fossil fuels, and evolutionary psychology can indicate the behavioral problems that make climate change hard to address, yet neither can provide the kind of self-reflection enabled by narrative.
>
> (53)

Nixon writes of Maathai's seed planting as an "act of intergenerational optimism" (2011, 134); likewise, *How Beautiful We Were* might be said to serve a similar purpose in its being a memento for future generations, as well as a source of inspiration. Although the characters' children chuckle at their parents' questions and worries, their grandchildren may decide that justice needs to be readdressed and that it is still worth saving Kosawa, a "splendid piece of Earth" (Mbue 2021, 359), from total destruction and abandonment.

References

Auty, Richard M. 1993. *Sustaining Development in Mineral Economies: The Resource Curse Thesis*. London and New York: Routledge.

Balkan, Stacey, and Swaralipi Nandi, eds. 2021. *Oil Fictions: World Literature and Our Contemporary Petrosphere*. Philadelphia, PHI: Pennsylvania State University Press.

Boyd, William. 1995. "Introduction." In *A Month and a Day: A Detention Diary*, edited by Ken Saro-Wiwa, vii–xiv. London and New York: Penguin Books.
Brown, A. L. 2014. "No Promised Land: The Shared Legacy of the Castle Bravo Nuclear Test." *Arms Control Today* 44 (2): 40–44.
Bruederle, Anna, and Roland Hodler. 2019. "Effect of Oil Spills on Infant Mortality in Nigeria." *Proceedings of the National Academy of Sciences* 116 (12): 5467–5471.
Climate Accountability Institute. 2020. "Carbon Majors." *Climate Accountability Institute* https://climateaccountability.org/carbon-majors/
Climate Accountability Institute. 2022. "Home." *Climate Accountability Institute.* https://climateaccountability.org
CNN. 2021. "Read the Latest Draft COP26 Agreement in Full." *CNN*, November 12. https://edition.cnn.com/2021/11/12/world/full-second-draft-cop26-agreement-climate-intl/index.html
Dwivedi, Om Prakash. 2023. "Subaltern Ecologies: Cultures of Concealment and Carbon Socialism." *Postcolonial Text* 18 (1&2): 1–16.
Deckard, Sharae. 2021. "Gendering Petrofiction: Energy, Imperialism and Social Reproduction." In *Oil Fictions: World Literature and Our Contemporary Petrosphere*, edited by Stacey Balkan and Swaralipi Nandi, 41–58. Philadelphia: Pennsylvania State University Press.
Gayle, Damien. 2022. "Just Stop Oil Campaigners Glue Themselves to Da Vinci Copy in Royal Academy." *The Guardian*, July 5. https://www.theguardian.com/artanddesign/2022/jul/05/just-stop-oil-campaigners-glue-themselves-to-da-vinci-copy-in-royal-academy
Ghosh, Amitav. 1992. "Petrofiction: the Oil Encounter and the Novel." *The New Republic*, March 2, 29–34.
Grasso, Marco, and Stefano Vergine. 2020. *Tutte le Colpe dei Petrolieri* [All the Fault of Oil Companies]. Milano: Mondadori Libri. Kindle edition.
Griffin, Paul. 2017. "The Carbon Majors Database." *CDP Carbon Majors Report.* https://cdn.cdp.net/cdp-production/cms/reports/documents/000/002/327/original/Carbon-Majors-Report-2017.pdf
Grubb, M., Cukwumerije. Okereke, Jun Arima, Valentina Bosetti, Ying Chen, James A. Edmonds, Shrrekant Gupta, Alexandre Köberle, Snorre Kverndokk, Arunima Malik, and Linda Yanti Sulistiawati. 2022."Introduction and Framing." In *IPCC, 2022: Climate Change 2022: Mitigation of Climate Change,* edited by Priyadarshni R. Shukla, Jim Skea, Raphael Slade, Alaa Al Khourdajie, Reéne van Diemen, David McCollum, Minal Pathak, Some, P. Vyas, Roger Fradera, Malek Belkacemi, Apoorva Hasija, Sigourney Luz, and Julliete Malley, 149–213. Cambridge and New York: Cambridge University Press.
Integrated Pollution Prevention and Control (IPCC). 1992. *Climate Change: The IPCC 1990 and 1992 Assessments.* Accessed January 29, 2022. https://www.ipcc.ch/site/assets/uploads/2018/05/ipcc_90_92_assessments_far_full_report.pdf
Mbue, Imbolo. 2021. *How Beautiful We Were.* New York: Penguin Random House. Kindle edition.
Munif, Abdulrahman. (1984) 1989. *Cities of Salt.* Trans. Peter Theroux. New York: Vintage.

Nixon, Rob. 2011. *Slow Violence and the Environmentalism of the Poor.* Cambridge, MA and London: Harvard University Press.
Okonta, Ike, and Oronto Douglas. 2001. *Where Vultures Feast: Shell, Human Rights, and Oil in the Niger Delta.* New York: Sierra Club Books.
Pandey, Manish. 2022. "Just Stop Oil: Why Protesters are Tying Themselves to Goalposts." *BBC News*, March 21. https://www.bbc.com/news/newsbeat-60795041
Perry, Alfred M., Kamal J. Araj, William Fulkerson, David J. Rose, Marvin M. Miller, and Ralph M. Rotty. 1982. "Energy Supply and Demand Implications of CO2." *Energy* 7 (12): 991–1004.
Riley, Tess. 2017. "Just 100 Companies Responsible for 71% of Global Emissions, Study Says." *The Guardian*, July 10. https://www.theguardian.com/sustainable-business/2017/jul/10/100-fossil-fuel-companies-investors-responsible-71-global-emissions-cdp-study-climate-change
Saro-Wiwa, Ken. 1992. *Genocide in Nigeria: The Ogoni Tragedy.* London, Lagos, Port Harcourt: Saros International Publishers.
Solnit, Rebecca. 2016. *Hope in the Dark: Untold Histories, Wild Possibilities.* Chicago, IL: Haymarket Books.
Szeman, Imre. 2012. "Introduction to Focus: Petrofictions." *American Book Review* 33 (3): 3–3.
Szeman, Imre, and Dominic Boyer, eds. 2017. *Energy Humanities: An Anthology.* Baltimore, MD: Johns Hopkins University Press.
The Center for Fiction. 2021. "Imbolo Mbue Launches *How Beautiful We Were* with David Ebershoff." *YouTube*, March 9. https://www.youtube.com/watch?v=6uPh9x2cYdk
Trexler, Adam. 2015. *Anthropocene Fictions: the Novel in a Time of Climate Change.* Charlottesville, VA and London: University of Virginia Press.
US Energy Information Administration. n.d. "U.S. Field Production of Crude Oil." *U.S. Energy Information Administration.* Accessed January 24, 2022. https://www.eia.gov/dnav/pet/hist/LeafHandler.ashx?n=PET&s=MCRFPUS2&f=M
United Nations Framework Convention on Climate Change (UNFCCC). 2021. "Glasgow Climate Pact." *UNFCCC.* https://unfccc.int/sites/default/files/resource/cop26_auv_2f_cover_decision.pdf
Vermeulen, Pieter. 2020. *Literature and the Anthropocene.* Oxon and New York: Routledge.
Watts, Michael. 2008. "Sweet and Sour." In *Curse of the Black Gold: 50 Years of Oil in the Niger Delta*, edited by Ed. Kashi, 37–62. New York: Power House Books.
Worland, Justin. 2020. "The Reason Fossil Fuel Companies Are Finally Reckoning With Climate Change." *Time*, January 16. https://time.com/5766188/shell-oil-companies-fossil-fuels-climate-change/
Zabala, Santiago. 2019. *Why Only Art Can Save Us.* New York: Columbia University Press.

6 Planetary precarity in performance ecopoetry

Poems to solve the climate crisis?

Jan Rupp

Introduction: planetary precarity, climate action and poetic praxis

In *The Great Derangement: Climate Change and the Unthinkable*, his long essay meditating on climate change as a challenge to the cultural imagination, Amitav Ghosh (2016) pointedly notes "global warming's resistance to the arts" (73). This resistance, for Ghosh, exacerbates the planetary climate crisis, leaving mankind with no appropriate cultural tools to fathom and attend to the precarious status quo, epitomized by a realization of impending man-made climate apocalypse in the age of the Anthropocene. Ghosh takes particular issue with the western realist novel, which he associates with a narrow focus on human scales as well as an individualistic outlook, both of which fall short of gauging the complex spatiotemporal trajectory of global warming and the collective response it requires. The plethora of so-called climate change fiction, which has meanwhile emerged as a symptom of growing climate anxiety, is undermined, in Ghosh's view, by a deprivileging and sidelining of genre fiction in the literary system. A challenge of existential magnitude like global warming deserves serious concern and mainstream attention, he insists, calling for the novel's cultural authority and literary fiction rather than speculative genre fiction.

While Ghosh's observation of mankind sleepwalking into climate doom is ominously plausible, his "provocative if at times simplistic" (Clark 2019, 97) account of the novel and other genres of climate change has met with some controversial debate.[1] One obvious blind spot of his argument is his somewhat one-sided account of the novel, which ultimately deemphasizes genre writing (including his own), as well as other traditions of storytelling, especially from the Global South, that do go some way towards capturing the scalar complexity of the Anthropocene and often help to imagine a habitable planetary future beyond the human. While Ghosh is right to point out representational challenges – "global warming's resistance to the arts" (2016, 73), that is – the critical engagement

DOI: 10.4324/9781003383598-9

and "resistance" mounted to man-made climate change by the arts can hardly be overlooked.[2] A major art form or genre that has not been given due attention yet, I argue, is ecopoetry, specifically of the kind circulating around international climate action. This poetry tends to mobilize a mode of performance, affect, and the spoken word, with the poet figuring as a moral messenger or cultural ambassador from some of the regions hardest hit by climate change. Staging a direct encounter between the poet or speaker and their audience, what I call "performance ecopoetry" renders concrete and personal what may seem like an abstract or distant process of climate change. Moreover, performance ecopoetry capitalizes on the way in which poetry, arguably far more than other genres, lends itself to the shifting media landscape of the 21st-century world. Rather than being published and archived as written texts only, many of the poems in question unfold their power as live performances or as multimedia video clips, being viewed and shared or even going viral online.

In this chapter, I set out to map and assess critically the generic and media affordances of performance ecopoetry, which, I suggest, contribute significantly to creating a popular mainstream platform, perhaps even a global public sphere, for tackling the challenge of climate change. In the chapter's two main sections to follow, I will combine a theoretical explication of performance ecopoetry with a close reading of poems that have emerged in the context of international climate action. My central case study focuses on the much anticipated (at the time of undertaking this research) 26th Conference of the Parties (COP) climate summit in Glasgow in November 2021 – COP26. The run-up to the summit featured a writing challenge entitled "Poems to Solve the Climate Crisis", which I have taken up in the title of my chapter. It will certainly take more than poems alone to solve the climate crisis, but what does it take for a poem, if not to solve, then at least to contribute to solving the climate crisis? In exploring this question, I will discuss a first example next, as well as fill in some historical and theoretical context, starting from the observation that recent years have seen a new poetic praxis of climate action, in sync with a wider revival of occasional, spoken word or performance poetry in the 21st century.

Pragmatics and poetics of performance ecopoetry

Tellingly, the first speaker to take to the stage at COP26 was "Instapoet" Yrsa Daley-Ward (2021). In her specially commissioned poem "EARTH TO COP", available both as a YouTube clip of her live performance (UN Climate Change 2021) and as an electronic publication in the style and uppercase aesthetic of her poetry on Instagram, she announced to the auditorium: "THE HOUR THAT YOU STAND IN/IS GOLDENING, ALREADY/COPPERING, ALREADY FIRE./ANCESTORS OF

TOMORROW./THIS IS AN INTERVENTION!/EARTH IS TALKING TO YOU/TODAY" (Daley-Ward 2021, 1). By picturing the earth catching fire, Daley-Ward not only offered a drastic metaphor of global heating, but also indexed the increasing frequency of catastrophic wild fires as a tangible effect of climate change.

Her poem echoed the doomsday narrative of COP26, which was widely billed as the "last, best moment" to take climate action. Another slogan at the event said: "We can do this, if we act now".[3] Daley-Ward chimed in with this message of hope and optimism, of saving the planet even at the last minute. Reminding delegates that "YOU WERE BORN/TO BE FRIENDS OF THE/PRECARIOUS GROUND" (2021, 1), she urged them "TO LEAD/WITH LIGHT" (1) and "BE SHINING FINAL/HOPES, DESIGNERS OF CHANCE/AND CHANGE" (2). However, her poem offered a powerful antidote to the iconic image of Blue Marble, which was omnipresent at the summit venue. Blue Marble is the famous photograph taken on the Apollo 17 space mission in 1972, showing the water-blue, life-giving planet Earth against the backdrop of a vast, black void of outer space. While drawing attention to the Earth's fragility, Blue Marble may be seen as problematic for glossing over the actual damage that the planet has already suffered. It is an ambivalent representation, susceptible of feeding a comforting sense of environmentalist nostalgia.[4] By contrast, Daley-Ward presents a very different blue or in fact red-blue planet: "THE HOUR IS/CRIMSON. TURNING, TURNING,/THE HOUR IS BRUISING/BLUEING, BLUE" (2). Rather than reproducing a detached view of fragile beauty set in timeless marble stone, Daley-Ward references the accelerating temporality of global warming and translates it into a visceral image of the bruised and hurting planet-body, where the Earth's blueness might only be glimpsed as a gradual and relative recovery or healing, if at all. Juxtaposed with Blue Marble, which hung as a large three-dimensional replica from the ceiling at COP26, this metamorphosing colour symbolism introduces a shocking twist and paradigmatically shows, in combination with Daley-Ward's dramatic delivery, the potential effects, "subversive possibilities" (Miller 2021, 170), and affective power of performance ecopoetry.

Daley-Ward's "EARTH TO COP" was not the only poem written for, and performed at COP26. The run-up to the summit also featured the aforementioned writing challenge "Poems to Solve the Climate Crisis", and a selection of the winners performed their poems at a fringe session in Glasgow, albeit to a much smaller audience than Daley-Ward's. In fact, poetry has seen a remarkable renaissance overall in recent years: "Poetry sales have skyrocketed in the past decade. The genre has become one of the fastest-growing categories in the publishing industry" (Ardam 2022, 2). This success arguably owes to poetry's "easy mobility" (Chasar 2022, 4), which, as "one of the most overlooked and understudied characteristics of poetry as a literary form" with a "rich and almost equally long history of being remediated and transmediated" (4), manifests itself once more in the 21st-century media world. Its easy transition to new digital and social media makes poetry well-placed to speak to a wide range of pressing topical issues, to serve as "protest literature" (Keown 2017, 930) and explore "poetics and aesthetics as a form of activism" (Huang 2022, 261). Not least, the conspicuous media presence of poetry and a new generation of poets are characterized by a degree of publicity afforded by few other genres.

In the wake of her much-noted inauguration poem (of Joe Biden in Washington, DC, on January 20, 2021) "The Hill We Climb", Amanda Gorman has also been championed for her earlier poem "Earthrise" (cf. Climate Reality 2018), dedicated to Al Gore's climate activism. Elsewhere, Kathy Jetnil-Kijiner has become a prominent voice in international climate action over the past decade or so. "Tell Them", written in 2011 and performed to much acclaim at the London 2012 Poetry Parnassus, tells of planetary precarity and environmental injustice in the South Pacific, a region at double risk from rising sea-water levels and toxic nuclear waste (Starr 2018). Vowing to fight for the survival of her home, Jetnil-Kijiner performed "Dear Matafele Peinem", a poem addressed to her baby daughter, at a United Nations (UN) climate summit in New York in 2014. She has since continued her writer-activism, drawing on her personal experience of the Marshall Islands to highlight the communal challenges of climate change as a global public affair.

A formative mode already in antiquity and in many oral literary traditions until today, such public poetry has often been discounted on aesthetic grounds. Yet, its combination of subjective lyric expression and communal address holds a distinctive potential for a range of collective challenges and proliferating crises in the 21st century. As a related case to global warming, refugee and climate-induced migration has become a prominent theme of spoken word poetry. A much-discussed poem in this connection is Warsan Shire's (2011) "Conversations about Home (at the Detention Centre)", which, in the wake of the so-called "refugee crisis" in Europe in 2015 and after, "was and is still being used as an antidote to the growing representation of refugees as spectacle and threat, as it gives voice to those refugee subjects made silent and inhuman by Western paranoia and xenophobia" (Leetsch 2021, 209). A significant part of the poem's success, including of a later (audio) version simply entitled "Home",[5] and read by the author herself, has been its circulation in a digital "space of reaction and response that is much more open and fluid than traditional means of publishing and disseminating literary works" (226). This example is in line with some general, overarching trends of recent poetic practice:

> In the twenty-first century, the internet has become a space for young poets to reach a global, connected audience – one that is on the one hand

> exceptionally attuned to their subject matters […], and on the other hand one that is generated by the digital equivalent of "word-of-mouth", that is, links, shares, likes and various algorithms.
>
> (226)

Indeed, it is the multimodal texture and the popular, social media appeal of many recent poems that certainly add to creating a new public sphere for tackling issues such as refugee migration and climate change.

Thanks to these media affordances, as well as to several pragmatic and poetic protocols specific to the genre, performance poetry clearly stands to provide an influential platform to rally support for climate action, with the digital space promising a global reach: "The barriers to global audiences are lower. The potential audiences are seemingly larger than ever. The most relevant or prominent platforms can change with remarkable speed" (Chasar 2022, 182). While this might benefit the cause of climate action, writers exploring environmental concerns will invariably have to confront the complex legacy of the poetic imagination, including lingering Romantic conceptions of nature, which makes for productive though not uncomplicated uses of poetry in the face of the planet's growing environmental precarity. As Timothy Clark (2019) has demonstrated, ecopoetry comprises a highly diverse body of work (57–77). While there is no set definition, ecopoetry emerged as a term around the turn of the 21st century, referring to texts that attend to accelerating environmental change in what began to be discussed as the Anthropocene. Linked to contemporary developments of environmental change, ecopoetry has also been traced in literary history, as in early ecocritical readings of Romantic poetry as accounts of recuperative nature for modern man. A more expansive and a more specific understanding of ecopoetry thus co-exist, allowing for tracing continuities and differences.

Current ecopoetry circulating around international climate action tends to be a subset of performance poetry, as I have argued, characterized by a specific array of pragmatic and poetic structures. While it may seem redundant to point out the pragmatics of performance, it is worth noting Elleke Boehmer's (2018) more fundamental reappraisal of the interplay between poetics, reading, and reception. Shifting attention from questions of representation to interpretation, Boehmer understands poetic structures as a "score for reading" (1–17). This

> idea of poetics as a score, as in music or dance, […] sees the reader as the interpreter of that score, even as its performer. As we read we negotiate actively with the text's latent and emergent properties, […] to the end of assembling and asserting our interpretation.
>
> (9)

Like a musical score, literary texts thus unfold fully only through the recipient's interpretive activities, their cognitive and affective engagement.

This fundamental dynamic is clearly replicated or even multiplied in the performance ecopoetry of international climate action. By definition, this poetic praxis is constituted as text and performance in one, acted out in tandem by the poet and their audience, and pre-figured by the text's poetic structures. Jetnil-Kijiner's "Tell Them" (2011) is a case in point. Programmatically, the imperative mood in the title signposts the poem's communicative set-up and rhetoric, centred around a message that is supposed to convey the precarious environmental reality of the Marshall Islands. In fact, the poem is a message, quite literally, written in the form of a letter that is enclosed, together with earrings and small baskets made from island materials, in a package that the speaker prepares for "my friend in the states" (Jetnil-Kijiner 2011, n.p.). The letter explains the provenance of these objects and asks the recipients (the speaker's friends and, by extension, the reader) to keep them as metonyms of a precious yet endangered island world and warn of the existential risks the Marshall Islands are facing: "*Wear these earrings/to parties/to your classes and meetings/to the grocery store, to the corner store/and while riding the bus*" (n.p.; emphasis in original). In Jetnil-Kijiner's video performance of the poem, recorded for the London 2012 Poetry Parnassus, this sense of urgency is further dramatized through the dark ambient visuality of a Marshall Islands background setting and the emotional tone of her voice (cf. Studio Revolt 2012).

A central motif echoing throughout, the poem's iterations of "tell them" remind the speaker of her role as a cultural go-between. The imperative mood is both a form of self-address and of addressing her readers or listeners, inviting them to share in her locally grounded experience and to pass it on. Thus, the act of telling and of (imaginatively) inhabiting this story is effectively multiplied, shared by the poet or speaker and passed on in turn by word of mouth among shifting audiences. As Boehmer argues, "shift[ing] the reader between and across different worlds, whether cultural, cognitive, or national" (2018, 13) is a special potential of postcolonial poetics, its "capacity to keep reimagining and refreshing how we understand ourselves in relation to the world and to some of the most pressing issues of our time" (1). Significantly, Jetnil-Kijiner's South Pacific world, as one of the regions hardest hit by global climate change, can be considered as emblematic of planetary precarity at large. In *Allegories of the Anthropocene*, Elizabeth DeLoughrey (2019) reads Jetnil-Kijiner's "Tell Them" as a quintessential example of island narratives from the Global South that allegorically manage to capture the spatiotemporal complexity of the Anthropocene.

In her study, DeLoughrey starts out by criticizing the fact that scholarship of the "Anthropocene cultural imaginary is focused almost exclusively on the viewpoints of the global north" (2019, 2). A "lack of engagement with postcolonial and Indigenous perspectives", she observes, has led to a sense of "novelty of crisis rather than being attentive to the historical continuity of dispossession and disaster caused by empire" (2). Consequently, DeLoughrey reroutes attention to regions in the Global South, which "have been and continue to be at the forefront of ecologically devastating climate change" (2). More specifically, she focuses on island narratives from the South Pacific to the Caribbean, covering a wide range of texts from Jetnil-Kijiner's poetry to contemporary Caribbean fiction. In all these examples, DeLoughrey notes, the island serves as an allegorical figure of "a world in ecological crisis" (2). Global South narratives, thus, mobilize a longer cultural history in which the island has been figured as "an analogue for the globe" (6). In the current climate crisis, this metonymic relationship allows for a "scalar telescoping between local and global, island and Earth" (6). Placing Jetnil-Kijiner's "Tell Them" in this context of Anthropocene time and space, we can ultimately see the poem as employing a rhetoric of communal address, binding the speaker and her audience into a collective effort for the sake of the planet's future.

If performance ecopoetry may hope to afford a larger pragmatics of climate action, it bears repeating that it can also build on distinctive potentials of representation. Poetry has variously been found to be well-placed as a genre of climate change, often in contradistinction to the novel. As discussed above, for Amitav Ghosh, the novel's legacy of realism and bourgeois life falls short of accommodating the disruptive and sudden impact of climate disaster, whereas "poetry, on the other hand, has long had an intimate relationship with climatic events" (Ghosh 2016, 26). Clark argues that poetry can be rendered sensitive to non-human actors and cataclysmic environments more easily:

> The true complexity of environmental issues has been perhaps easier to represent in new or revised forms of poetic practice than in prose forms like the novel, short story or the non-fiction essay, or in theatre. One advantage of poetry is that its removal from the conventional constraints of prose narrative renders it more open to representing multiplicity, and even contradiction and indeterminacy, and to do so without its readers necessarily feeling the lack of some clear storyline.
>
> (29)

As opposed to a concern with individual lives and human lifespans in the (realist) novel, poetry's more open and potentially multiperspectival form is indeed able to represent and grasp scalar complexity, as Jetnil-Kijiner's "Tell Them" may go on; to show once more. With its structure of address and its performance, by video, to audiences in the Global North, the poem traverses vast spatial distances, underscoring the significance of climate change in the South Pacific for the planet as a whole. Similarly, the poem travels widely in time, straddling two ends of a temporal continuum by telling a creation story of the Marshall Islands alongside an apocalyptic narrative of human intervention and impending disaster:

> *tell them our islands were dropped*
> *from a basket*
> *carried by a giant*
> *tell them we are the hollow hulls*
> *of canoes as fast as the wind*
> *slicing through the pacific sea*
> *[…]*
> *and after all this*
> *tell them about the water*
> *how we have seen it rising*
> *[…]*
> *tell them what it's like*
> *to see the entire ocean__level__with the land*
> *tell them*
> *we are afraid*
>
> (Jetnil-Kijiner 2011, n.p.; emphasis in original)

What emerges from this wide-ranging spatiotemporal account is a precarious yet symbiotic multiplicity, an intricate and entangled human-environment relationship ("*we are the hollow hulls*", "*we are the ocean*") that is central to the poem's design to prompt action.

COP26: poems to solve the climate crisis?

Taking up again my earlier reading of COP26 poetry and segueing to some further examples, it is interesting to see whether the "Poems to Solve the Climate Crisis", submitted to the eponymous writing competition in the run-up to the climate summit in Glasgow, live up to the challenge, and how they compare to the successful performance ecopoetry by Daley-Ward, Jetnil-Kijiner and others. Created by the Poetry Society's Young Poets Network, and set and judged by the poet Louisa Adjoa Parker, the competition was indeed directed at younger writers, though many of the prize-winning candidates were already published poets. As per the call, the challenge invited participants "to respond to the UK's COP26 Presidency theme of Nature" and "to imagine solutions to the climate crisis inspired by the natural world".[6] These references to nature and the natural world resonate with the agenda of People Need Nature, a UK-based charity which co-organized the challenge. On its website, the charity presents a list of mottos, illustrated by photographs of open woodland and green meadows – aphorisms like "Nature Doesn't Need People, But People Need Nature" or "Nature provides us with solace and a place to escape from difficult times".[7] While these statements sound plausible enough, they arguably recall "inherited Romantic assumptions that pit human affairs against a would-be 'sacralised' nature" (Clark 2019, 59). They not only fall back on dichotomies long deconstructed in ecocritical thinking, but also tend to ignore the fact that nature has already been irretrievably spoilt by human intervention.[8] Poetry based on such escapist nostalgia would seem to hold little promise for helping to prevent further climate change.

The submissions for the "Poems to Solve the Climate Crisis" competition navigate these premises and the overall challenges of ecopoetry in a broad range of different ways and poetic styles. They have all been published as written texts and video or audio clips online, with some of the prize-winning poems featuring in a live session at COP26. While drawing on an oral, dramatized style to different degrees, the poems were all designed for a mode of performance rather than silent reading. Between them, the first two prize-winning texts demonstrate the broad range of submissions and solutions offered to the climate crisis "An Alternative Geometry of the Universe" by Maggie Wang from the US and "A Positively Violent Poem in Five Parts" by Jayant Kashyap from India show very different outlooks, as well as a variety of formal and conceptual choices, respectively. Wang's poem has a very regular form, consisting of nine two-line stanzas. It is accompanied by an audio clip of the author's reading only, which adds to the text's calm and meditative quality. The poem opens with the speaker and another character eating cherries in summer, but it soon zooms out from the human habitat to contemplate cycles of day and night, shifting weather and the seasons:

> In June, Ba buys cherries, and we run our fingers along the skins,
> which gleam like piano keys or the moon after fog.
> We eat the cherries on the back porch,
> watching the maples turn the last of the sunlight over their leaves.
> Above, the clouds are gathering again,
> and the birds are tracing fractals along their edges.
>
> (Wang 2021, n.p.)

Within that natural rhythm, a self-sufficient economy of salmon, bears and bees communes. Any sense of a human presence is lost, save for a new summer of ripening cherries at the end. As Wang comments in her recording:

> This poem was an attempt to envision what the earth might look like without us, or what it might look like if we took a step back and recognized our place as one link or one stop in the broader cycles of the natural world.
>
> (Wang 2021, n.p.)

Wang certainly succeeds in presenting an alternative geometry of the universe, a biocentric world to be gazed at by humans as silent bystanders. Her soothing voice and the poem's many sensory images offer the reader a concentrated and comforting experience of a greater whole. This alternative geometry might be suggestive of the late James Lovelock's (1988) Gaia theory, of earth as a self-regulating system, or of the late Bruno Latour's (2017) reappraisal in *Facing Gaia: Eight Lectures of the New Climatic Regime*, in which Latour replaces the anthropocentric concept of the Anthropocene with the idea of multiple agency and rights distributed across a network of human as well as non-human actors and environments. However, in Wang's poem, the idea of human smallness, of cyclical time and the universe running its course, is also surprising in a context like COP26, because international climate action is premised on acknowledging the disproportionate extent of man-made climate change. Whether such an alternative conception will serve as a solution to the climate crisis – or let humanity off the hook by picturing nature as a noble savage – is clearly debatable.

Jayant Kashyap's "A Positively Violent Poem in Five Parts", which was voted second place, takes a very different approach in form and content. Where Wang's regular poetics enacts contemplation and stepping back, or almost amounts to a poetics of inaction, Kashyap's poem employs a fragmented and agitated style, staging the violent disruption that the title speaks of and projecting a sense of restless urgency. Kashyap takes the reader to an unnamed, generic tourist destination, a beach covered in plastics, which becomes the scene of rebellious climate action: "we noticed the beach had more plastics and whatnot/and not shells" (Kashyap 2021, n.p.). Uncannily, plastic is a pervasive symbol of toxic pollution and disease in this scene, affecting humans and animals alike: "even the castles the kids built near water/weren't pure sand anymore/sometimes the bottlecaps serve as lighthouse roofs/sometimes the straws [...] as catheters/for turtles" (n.p.). The poem's setting, marked by the slow violence of waste and environmental degradation, is in turn subjected to the "positive violence" of climate activism in the style of Extinction Rebellion or similar groupings.

The poem's we-narrative tells of a group of young locals who dismantle the tourist infrastructure, such as swimming pools and hotels. In turn, they plant trees ("because later the trees could oxygenate water", n.p.) and clean the beach, all in an attempt to try and transform the place back to a state of nurturing nature ("that day we left it like nature", n.p.). While their violence is characterized as positive, it is challenged by the voice of an older generation, who is equally represented by the poem's disjointed, multiperspectival account. Moreover, the young activists seem to reflect on their rebellion in a rather nuanced way, as their opening self-characterization suggests: "*That day we ruined the swimming pool*/everyone was a hypocrite and angry/we pulled out the tiles with bare hands and nails/our knuckles bled quietly into tattoos" (n.p.; emphasis in original). Right from the start, Kashyap's poem pre-empts any idea of easy solutions. Even the young activists must realize that they are implicated in a complex web of human-nature relationality, where replacing swimming pools with trees will not simply restore nature and climate action must be negotiated collectively: "*Next morning we were met by a racket of people*/wanting to talk [...] having a mind to argue/swear for generations" (n.p.; emphasis in original). While, at first, some of the older community members are "prepared to fight/kill someone saying *son it doesn't matter who –/ or what– you are*", support for the young climate activists ultimately grows: "*And that evening we cleaned the beach*/there were more people with us now than against" (n.p.; emphasis in original). As text and performance, this negotiation and collaboration of climate activism – importantly extending beyond the page as the reader navigates a fragmented poetic score – is a process that Kashyap's poem helps engender and move forward, capitalizing on the pragmatics and poetics of performance ecopoetry.

Conclusion: performance and climate change

As these close readings demonstrate, along with my earlier discussion of the texts by Yrsa Daley-Ward and Kathy Jetnil-Kijiner, performance ecopoetry holds a particular promise to inspire climate action and ensure the planet's future habitability. The dual, personal, and intimate as well as public character of performance poetry may render abstract notions of planetary change concrete. Together with its open form and elastic scope of representation, the genre conveys local experiences as metonyms of global developments to potentially distant audiences. Moreover, the oral nature of this poetic praxis, capitalizing on the spoken voice for presence effects and affect, as well as the multimodal texture, popular appeal, and social media use of many recent poems all combine and work towards creating a global public sphere to attend to the planet's environmental precarity. Its easy transition to a context of new media and 21st-century attention economies is another factor in setting poetry apart from other genres and in distinguishing it as a medium of tackling climate change.

Events like COP26 no doubt benefit from the media praxis of performance poetry and other arts, as a direct, communal and affective form of address. Conversely, international climate action has been instrumental in providing an alternative platform for poetry as a body of ecologically inflected world writing. At a time when the world literary system continues to revolve around the novel, this alternative circuit and genre of world literature deserves still greater attention. Acted out in the co-presence of speaker and audience and designed to mobilize collective action, the poems' primary mode of – medialized – performance ultimately requires an extension of academic practices in order to accommodate this media praxis of ecopoetic world writing.

As my close readings have also shown, the potential of ecopoetry to inspire a pragmatics of climate action is closely bound up with the question of precisely what concepts of environment are mediated, and in which poetic form. Here, nostalgia for unspoilt nature, Romantic poetic tropes, and a pragmatics of contemplation, or such popular images in western environmentalism as Blue Marble, would seem to prevent rather than facilitate meaningful solutions to the climate crisis. Indeed, to move beyond the mixed success and criticisms of events like COP26 as "a Global North greenwash festival",[9] climate action, poetic praxis, and academic practices are well advised to place still greater emphasis on accounts of Anthropocene time and space from the Global South.

Notes

1 See Adeline Johns-Putra (2016), who notes "a recent emergence of climate change-related novels by 'serious' or 'literary' authors, who tend to be published by well-known presses and receive mainstream media attention" (267).

2 As for literature, cf. Adam Trexler and Johns-Putra (2011) as well as the broad spectrum of genres discussed in Johns-Putra, including "climate change in drama and poetry, where some significant developments have occurred" (2016, 266). However, Johns-Putra does not deal with the specific case of ecopoetry in the context of international climate action. See also Chao Xie (2022), who notes a relative paucity of scholarship on climate change poetry compared to climate change fiction.

3 The latter slogan was part of the stage design at COP26, the former a set phrase in news reporting and political statements about the event.

4 On (criticisms of) Blue Marble as a planetary imaginary, see T.J. Demos (2017, 19–23).

5 See https://www.youtube.com/watch?v=nI9D92Xiygo (accessed Mar. 10, 2023) for the author's reading, in a revised version of her original poem (cf. Shire 2011).

6 See this webpage of the Young Poets Network, featuring the writing challenge: https://ypn.poetrysociety.org.uk/workshop/poems-to-solve-the-climate-crisis-a-new-challenge-with-people-need-nature-louisa-adjoa-parker/ (accessed Jan. 15, 2023).

7 See the charity's homepage: https://peopleneednature.org.uk (accessed Jan. 15, 2023).
8 On this critique see inter alia, Timothy Morton (2009).
9 This was a widely shared tweet by Greta Thunberg (quoted in Brooks 2021), which was echoed by the largely critical news coverage of COP26.

References

Ardam, Jacquelyn. 2022. *Avidly Reads Poetry*. New York: New York University Press.

Boehmer, Elleke. 2018. *Postcolonial Poetics: 21st-Century Critical Readings*. Basingstoke: Palgrave Macmillan.

Brooks, Libby. 2021. "The Good, the Bad and the Irn-Bru: How Cop26 Played Out." *The Guardian*, November 12. https://www.theguardian.com/environment/2021/nov/12/the-good-bad-irn-bru-how-cop26-played-out

Chasar, Mike. 2022. *Poetry Unbound: Poems and New Media from the Magic Lantern to Instagram*. New York: Columbia University Press.

Clark, Timothy. 2019. *The Value of Ecocriticism*. Cambridge: Cambridge University Press.

Climate Reality. 2018. "'Earthrise' by Amanda Gorman." https://www.youtube.com/watch?v=xwOvBv8RLmo

Daley-Ward, Yrsa. 2021. "EARTH TO COP." https://unfccc.int/documents/362687

DeLoughrey, Elizabeth M. 2019. *Allegories of the Anthropocene*. Durham, NC: Duke University Press.

Demos, T. J. 2017. *Against the Anthropocene: Visual Culture and Environment Today*. London: Sternberg Press.

Ghosh, Amitav. 2016. *The Great Derangement. Climate Change and the Unthinkable*. Chicago, IL: University of Chicago Press.

Gorman, Amanda. 2021. *"The Hill We Climb": An Inaugural Poem for the Country*. London: Viking.

Huang, Hsinya. 2022. "Climate Change and Indigenous Sovereignty in Pacific Islanders' Writing." In *The Cambridge Companion to Literature and Climate*, edited by Adeline Johns-Putra and Kelly Sultzbach, 257–268. Cambridge: Cambridge University Press.

Jetnil-Kijiner, Kathy. 2011. "Tell Them." https://jkijiner.wordpress.com/2011/04/13/tell-them

Jetnil-Kijiner, Kathy. 2014. "Dear Matafele Peinem." https://www.youtube.com/watch?v=DJuRjy9k7GA

Johns-Putra, Adeline. 2016. "Climate Change in Literature and Literary Studies: From Cli-Fi, Climate Change Theater and Ecopoetry to Ecocriticism and Climate Change Criticism." *WIREs Climate Change* 7 (2): 266–282.

Kashyap, Jayant. 2021. "A Positively Violent Poem in Five Parts." https://ypn.poetrysociety.org.uk/uncategorized/a-positively-violent-poem-in-five-parts

Keown, Michelle. 2017. "Children of Israel: US Military Imperialism and Marshallese Migration in the Poetry of Kathy Jentnil-Kijiner." *Interventions* 19 (7): 930–947.

Latour, Bruno. 2017. *Facing Gaia: Eight Lectures of the New Climatic Regime*. Translated by Catherine Porter. Medford, MA: Polity.
Leetsch, Jennifer. 2021. *Love and Space in Contemporary African Diasporic Women's Writing: Making Love, Making Worlds*. Cham: Palgrave Macmillan.
Lovelock, James E. 1988. *Gaia: A New Look at Life on Earth*. Oxford: Oxford University Press.
Miller, Alyson. 2021. "A Digital Revolution? Insiders, Outsiders, and the 'Disruptive Potential' of Instapoetry." *Arcadia* 5 (2): 161–182.
Morton, Timothy. 2009. *Ecology without Nature: Rethinking Environmental Aesthetics*. Cambridge, MA: Harvard University Press.
Shire, Warsan. 2011. "Conversations about Home (at the Detention Centre)." In *Teaching My Mother How to Give Birth*, 24–27. London: Flipped Eye Publishing.
Starr, Marlo. 2018. "Paradise and Apocalypse: Critiques of Nuclear Imperialism in Kathy Jetnil-Kijiner's *Iep Jāltok*." *Commonwealth Essays and Studies* 41 (1): 119–130.
Studio Revolt. 2012. "'Tell Them' by Kathy Jetnil-Kijiner at Southbank Centre for London 2012 Poetry Parnassus." https://www.youtube.com/watch?v=w9D88ST9qbw
Trexler, Adam, and Adeline Johns-Putra. 2011. "Climate Change in Literature and Literary Criticism." *WIREs Climate Change* 2 (2): 185–200.
UN Climate Change. 2021. "Yrsa Daley-Ward at the Opening of the #COP26 World Leaders Summit." https://www.youtube.com/watch?v=-3ilJn3g3vA
Wang, Maggie. 2021. "An Alternative Geometry of the Universe." https://ypn.poetrysociety.org.uk/uncategorized/an-alternative-geometry-of-the-universe
Xie, Chao. 2022. "Climate Change in Contemporary British and Irish Poetry and Poetic Criticism: Literary Representation and Environmental Activism." *WIREs Climate Change* 14 (1): 1–13.

7 Toward critical self-reflection and a vigilant sense of precarity

Why read pandemic literature during a pandemic

Scott Slovic

Introduction: the value of fiction in promoting critical self-reflection

In the spring of 2020, soon after people around the world began to change their lives in reaction to the Covid-19 pandemic, I started to think about the psychological responses to the public health crisis I had observed through the media and in my own life. At the top of the list I compiled for my June 2020 article for *Bifrost Online*'s special cluster of environmental humanities perspectives on Covid-19, "COVID World, COVID Mind: Toward a New Consciousness", was the idea that the incipient pandemic seemed to have stirred up a sense of universal vulnerability, a condition in which people of all walks of life had become fearful of an invisible and potentially deadly virus, resulting in extraordinarily careful behavior, including mask-wearing, maintaining social distance, avoiding public spaces, and even wiping down groceries with antiseptic swabs (Slovic 2020, n.p.). This level of mindfulness struck me as both strange and potentially useful, evidence that the public imagination in conditions of anxiety could be inspired to a much higher level of vigilance than normally exists – and that such vigilance could be translated into careful action as a way of mitigating sickness and death. Could this vigilance somehow translate to other aspects of our lives in the 21st century? Might we somehow learn to care about other crises, such as global climate change and mass extinction, as a result of information transmitted to us through various media? In this chapter, I explore several examples of recent American pandemic fiction in order to consider how these works – Geraldine Brooks's (2001) *Year of Wonders*, Peter Heller's ([2012] 2013) *The Dog Stars*, and Lawrence Wright's (2020a) *The End of October* – provide representations of pandemics or pandemic-like situations that confirm or deepen our lived experience of an actual pandemic, helping readers to think more critically about the meaning of Covid-19 and disease more generally.

I found myself describing the feeling of vulnerability that arose during the pandemic as a form of "ecoprecarity", using the term developed by

DOI: 10.4324/9781003383598-10

Pramod K. Nayar (2019) to capture the sense of danger resulting from certain exposures to the external world and the fragility of the world itself as a result of particular conditions created by human activity. "Precariousness", writes Nayar, "is the effect of an exposure to the world, which then inflicts injury" (2017, xviii). As he explains in his 2017 volume, *Bhopal's Ecological Gothic: Disaster, Precarity, and the Biopolitical Uncanny*, the precarity emerging from a particular event or condition occurs not only in an immediate physical way but also through the textual representation of the historical situation. In his book about the terrible Union Carbide chemical accident of 1984, in which methyl isocyanite leaked into the environment, killing 30,000 people immediately and injuring approximately half a million near the central Indian city of Bhopal, Nayar explains that the accident has continued to haunt society for nearly 40 years, dominating Bhopali lives and stimulating public concern about the dangers of industrial society and the frequently unethical practices of multinational corporations in externalizing the risks and costs of their activities. Nayar's study focuses on the Gothic quality of the art and literature that have emerged from the disaster, a quality that frequently results in an unsettling psychological affect, a feeling of disquiet. This psychological agitation is a positive thing, a way for disaster texts to enable the public to maintain a heightened sense of vigilance in the face of other potential threats. The cultural representation of vulnerable Bhopali lives helps to create heightened public awareness, not only with regard to industrial risks but in the broader context of environmental risk, including exposure to disease. While Nayar focuses specifically on the story of the Bhopal disaster and the subsequent representation of that disaster in his 2017 study, it is clear that he understands the lessons of "Bhopal's ecological Gothic" to be applicable more widely to what he calls "the culture of precarity" (2017, xviii). Nayar's 2019 study, *Ecoprecarity: Vulnerable Lives in Literature and Culture*, shows that he does not consider the Bhopal disaster to be an isolated incident but rather a notable example of a broader pattern of human and nonhuman precarity. In this chapter, I will argue that pandemic literature, particularly fiction with its sometimes graphic representation of human suffering, is a powerful example of the cultural representation of precarity.

In *Dancing with Disaster: Environmental Histories, Narratives, and Ethics for Perilous Times*, Kate Rigby (2015) offers a chapter devoted specifically to the literature of catastrophic disease titled "Spreading Pestilence", in which she notes that, unlike other forms of environmental catastrophe such as earthquakes and volcanoes, which are strictly natural, disease can have hybrid causes, including human agents of spreading. Rigby writes:

> pestilence spreads, to be sure, and over the past 150 years our understanding of the multiple other-than-human agencies responsible for the

> proliferation of infectious diseases has grown enormously; but so too have those sociocultural practices through which humans themselves inadvertently spread pestilence across the planet.
>
> (52)

She quotes Nigel Clark (2011), who argues that "In a world of intensifying global interconnectivity, we are multiplying vectors and niches for our microscopic nemeses far faster than we can physiologically or culturally adapt to their exertions" (28; quoted by Rigby 2015, 52). The particular emphasis in Rigby's environmental humanities approach to widespread disease is the idea that our own vulnerability to disease results from what she describes as "our utter dependence upon an unruly biosphere that was not designed for our exclusive benefit" (2015, 52). She paints a picture of coexistence and mutual responsibility, arguing that "reducing human susceptibility to contagion cannot be divorced from ethical questions concerning our treatment of other animals" (52–53).

Although Rigby does not offer a broad overview of pandemic literature, she does write at length about the novel that many consider to be the original text in the genre, at least within the western canon: Mary Shelley's 1826 work *The Last Man*. Novels such as this, according to Rigby, through their complex narrative structures, facilitate the capacity for "mindfulness" and "critical self-reflection" among readers (2015, 79). These may well be not the only fundamental reasons for reading any kind of literature, but more specifically the essential reasons for us to read literary texts that teach us about the kinds of experiences we are living through at any given moment: mindfulness and critical self-reflection. Although she does not offer a detailed explanation of what she means by mindfulness, it seems to me that Rigby is referring to the increased capacity of readers to understand their own mortality and their susceptibility to frailty and suffering when exposed not only to actual disasters, including pandemics, but also when exposed to cultural texts telling the stories of such disasters. The potential for readers to place themselves imaginatively in the experience of being at risk is what Rigby appears to be describing in her study.

It is interesting to think about Rigby's notion of mindfulness in relation to Molly Wallace's 2016 study of "risk criticism", which emphasizes the challenges of appreciating threats from man-made or natural events that have actually occurred versus events, such as full-scale nuclear war, that have not yet been experienced. Wallace argues that ecocriticism (textual analysis from an environmental perspective) is a way of reading that emphasizes engagement with "real", experiential phenomena. As she puts it, citing environmental critic Lawrence Buell (2005):

> Emphasizing the persistent "real" referent of the present and continuing environmental damage, ecocritics […] "looked to the movement

> as a way of 'rescuing' literature from the distantiations of reader from text and text from world". [...] One reads speculative fiction of nuclear holocaust in order to learn how to avoid it; one reads nature writing in order to learn how to be more ecological.
>
> (Wallace 2016, 9; quoting Buell 6)

I would extend this to suggest that one reads pandemic fiction in order to be more sensitive to a wide range of risky possibilities in contemporary life, not only the persistent threat of infectious diseases (even during non-pandemic times) but ordinary and extraordinary threats, ranging from street violence to the seemingly distant (but profoundly ominous) threats posed by global climate change, mass extinction, and spreading industrial toxicity. Wallace quotes Jacques Derrida's 1984 essay on nuclear war, "No Apocalypse, Not Now", to emphasize that environmental risk narratives are fundamentally different from what Derrida defines as the "fabulously textual" narratives of nuclear war, which describe entirely imagined (not experienced) scenarios (Derrida 1984, 23; quoted by Wallace 2016, 9). Likewise, the fact that medical crises and environmental degradation have actually been experienced by human beings and can be represented in cultural texts such as pandemic fiction, lends an immediacy and emotional intensity to the reading of such texts. The awareness that such events have happened in the past, are currently happening, and could happen again in the future, triggers what I am calling "mindfulness" in this chapter – a heightened sense of vigilance.

In Rigby's reading of Mary Shelley's novel from the early 19th century, she explains how critical self-reflection leads to an awareness of the need for humans to think of disease prevention not in a militaristic sense, as merely defeating alien and dangerous microbes, but rather to consider, as she puts it,

> disease prevention and treatment [...] along ecological lines as a multispecies project [... i.e.] conservation of wildlife habitat, the cessation of factory farming, protection of our own microbial messmates, and the stabilization of human population and consumption to levels that would allow for the continued flourishing of diverse other-than-human lives [...]. Such multispecies practices of disease mitigation and amelioration imply the acknowledgment of our transcorporeal connectedness with one another and other others, both in life and death.
>
> (Rigby 2015, 82–83)

Rigby concludes her discussion by referring to another classic of pandemic literature, Albert Camus's 1947 novel *The Plague*, in which he specifically criticizes human insularity and hubris, or as Rigby puts it, "complacent human self-enclosure" (83). Camus describes the townsfolk in his novel as being "wrapped up in themselves; [...] they were humanists; they disbelieved in pestilences" ([1947] 1948, 35). In other words, the illusion of our own isolation and superiority comes back to haunt us during pandemics, so we would be well served to keep this in mind to ward off future disease. This critique of human exceptionalism is particularly relevant for discussing Sarah McFarland's (2021) study of "eco-collapse narratives" including the disease narrative presented in Peter Heller's ([2012] 2013) novel, *The Dog Stars*, which will be considered later in this chapter.

Three novels: storying the psychological aspects of the pandemic experience

To turn to the pandemic novels mentioned at the beginning of this chapter, Geraldine Brooks's 2001 novel, *Year of Wonders*, is a historical narrative, set in mid-17th-century (1666) England, in a rural village. The first-person narrator, Anna, is a working-class woman, a housemaid, whose husband has fallen victim to the plague. Anna admits early in the narrative that she is feeling intensely isolated from other people, suffering each evening from "a silence so thick it falls upon me like a blanket. Of all the lonely moments of my day", she states, "this one is always the loneliest. I confess I have sometimes been reduced to muttering my thoughts aloud like a madwoman when the need for a human voice becomes too strong" (Brooks 2001, 9). This brings to mind the sense of intense loneliness that many people experienced during the Covid-19 pandemic, the feeling of extreme isolation and its powerful psychological effects.

Anna works as a housemaid in the home of the local preacher, Mr. Mompellion, and his wife. Another man named Mr. Viccars, a tailor, joins the household at one point, but soon contracts the disease that is raging through the village. Brooks presents Viccars's illness in vivid detail, describing how his head is

> pushed to the side by a lump the size of a newborn piglet, a great, shiny, yellow-purple knob of pulsing flesh [...] flushed scarlet, or rather, blotched, with shapes like rings of rose petals blooming under his skin. (2001, 42)

When Anna seeks to comfort Mr. Viccars, he exclaims, "For the love of God, Anna [...]. Get thee gone from here! Thou can't help me! Look to thyself!" (43). Despite Kate Rigby's notion that pestilence should inspire us to think of others and act in a spirit of generosity and mindfulness, *Year of Wonders* demonstrates just how strongly the sense of isolation and self-preservation prevails during pandemic experiences, even when there is not an active government lockdown in place.

Ironically, during a time of rampant death and isolation, Brooks's narrative also reveals that there are moments of intense community and even joy amidst the struggle. For instance, in one memorable scene in *Year of Wonders*, Anna delivers a baby for another townswoman. The description of the delivery is strangely akin to the gruesomeness of disease and death, full of blood and shouting: "When the little feet dangled at last through the womb opening", narrates Anna,

> the pace changed, and everything became urgent. I knew that the pulsing cord must on no account get crushed, and so with the greatest difficulty, I forced a hand past the buttocks and pushed it back. Mary screamed and shook with agony, and I felt scalding sweat run down my own back.
>
> (2001, 122)

Eventually the infant appears, "a small, slippery boy. And a moment later, he was yelling, too". "We laughed", says Anna. "And, for an hour, in that season of death, we celebrated a life" (123). To me, this also represents a key aspect of the pandemic experience, the idea that despite the many struggles, there are moments of ordinary joy. At the same time, it is a measured joy, one inevitably haunted by an awareness of the larger trauma. Brooks's narrator, whose own young children have died from the plague, remembers,

> even in the midst of that joy, I knew I would have to leave the babe at his mother's breast and return to my own cottage, silent and empty, where the only sound that would greet me would be the phantom echoes of my own boys' infant cries.
>
> (2001, 123)

Another aspect of *Year of Wonders* that rings true from the Covid-19 experience is the struggle of the local people in the village to find a cure for the disease. They could not expect help to come from outside of the village. There is a motif of self-reliance and scientific exploration as the characters seek to develop a herbal remedy. One townswoman says to Anna:

> For most, […] the onset of the disease spells the end of life. What we must do, here in this sorry little cottage, is find all the herbs of a strength-giving virtue and combine them in a tonic to fortify the healthy.
>
> (154)

The women page through ancient guides in Latin and Greek and try to match descriptions of various plants with the parts of the human body known to be afflicted by the disease. The narrator says:

> By afternoon, we had assembled the weapons for our armory. Nettle for the blood. Starwort and violet leaves for the lungs. Silverweed to cool a fever. Cress for the stomach. The worts of blow-ball for the liver, bat-weed for the glands, and vervain for the throat.
>
> (Brooks 2001, 154–155)

There is something strangely practical and soothing in the scientific approach to the plague, for Anna and the other characters. The conflict between rational science and superstitious religion is a prominent theme in Brooks's novel, and even in the instance of the contemporary Covid-19 pandemic, we can see some people more distracted with concerns about whether the disease is real or not and who may or may not be responsible for it than with actually taking practical steps to overcome the contagion. In one extended meditation, Brooks writes:

> Perhaps the Plague was neither of God nor the Devil, but simply a thing in Nature, as the stone on which we stub a toe.
>
> […] Did I really believe that God put the rock in my path to trip me? Some would say certainly: the finger of God places every speck of dust. I did not see it so. Yet I would have inclined to believe God's hand at work if, as a result of the rock, I'd struck my head and lay now fatally injured. So where, exactly, in the design of the world, did I believe that matters tilted the scale sufficient to garner God's notice? If I did not think He cared for the lie of a rock, why should I believe that He cared for a small life such as mine? It came to me then that we, all of us, spent a very great deal of time pondering these questions that, in the end, we could not answer. If we balanced the time we spent contemplating God, and why He afflicted us, with more thought as to how the Plague spread and poisoned our blood, then we might come nearer to saving our lives.
>
> While these thoughts were vexing, they brought with them also a chink of light. For if we could be allowed to see the Plague as a thing in Nature merely, we did not have to trouble about some grand celestial design that had to be contemplated before the disease would abate. We could simply work on it as a farmer might toil to rid his field of unwanted tare, knowing that when we found the tools and the method and the resolve, we would free ourselves, no matter if we were a village full of sinners or a host of saints.
>
> (215)

To me, this is an especially important passage as it highlights the idea that in order to save lives during a pandemic, science must precede religion and superstition, an idea that was frequently emphasized during the Covid-19 pandemic. *Year of Wonders* is set in a historical period of notable superstition (remember this is before the Age of Enlightenment in Europe) and relatively early in the understanding of infectious disease, as the science of vaccinations and other knowledge about how the spread of disease occurs and might be thwarted did not develop until the 18th century. As the novelist Brooks represents her characters, some of them, especially her uneducated female protagonist, were beginning to think scientifically about disease long before the social establishment moved in this direction. In a terrible irony, however, the novel also sheds light on the brutal way such women were treated in the 1600s, regarded not as scientific visionaries but as witches and threats to the community.

My next example, Peter Heller's *The Dog Stars* ([2012] 2013), is set in the late 20th century or early 21st century (it is not exactly clear when) in the USA. The novel represents the life of a central character named Hig, who has survived some kind of influenza epidemic that has decimated the population. The epidemic has occurred earlier, so the novel is set during the aftermath, when Hig lives alone with his dog, named Jasper, somewhere in the mountains of Colorado. Sarah McFarland's 2021 study *Ecocollapse Fiction and Cultures of Human Extinction* does a particularly good job of demonstrating how Heller's work disrupts the idea of human exceptionalism, creating an atmosphere of extreme precarity and susceptibility not only to disease but to other dangers, including the potential to be killed by other survivors who might be seeking resources. As McFarland argues, Heller's novel, like Cormac McCarthy's *The Road* (2006), demonstrates the fallacy of human exceptionalism, the idea that while the earth itself may be degraded by human activity, we ourselves will be spared the fate of suffering and extinction. In Heller's post-pandemic narrative, the world has been vastly depopulated of human beings. We do not see much of the world, only the region that Hig can reach by taking his small plane on reconnaissance flights across the vacant landscape of Colorado. As McFarland explains, the novel "explores the psychological effects of loss and individual adaptations to the breakdowns of human civilization and concomitant cultural transformations" (2021, 88).

After nine years of almost complete isolation, following the death of his beloved dog, Hig flies beyond the fuel capacity of his small plane. He thinks to himself: "I cannot live like this. Cannot live at all really. What was I doing? Nine years of pretending" (Heller [2012] 2013, 162). His desperate outreach for human contact introduces Hig to two more people, Pop and Cima, who accompany him back to the abandoned airport hangar where he lives. Gradually, though, Hig loses everyone he knows, and his narrative documents, as McFarland puts it, "the human species going extinct, individual by individual, in a rapidly warming world" (2021, 90). In addition to the narrative of Hig's travels and his efforts to stay away from marauding bands of dangerous fellow survivors (somewhat like what McCarthy describes in the post-apocalyptic scenario of *The Road*), we receive glimpses into the desolate cavern of Hig's consciousness. Heller writes, for instance:

> I stood in the shade of the tree in the cool breath of the moving water and let the sound, the light breeze blow through me. I was a shell. Empty. Put me to your ear and you would hear the distant rush of a ghost ocean. Just nothing. The slightest pressure of current or tide could push and roll me. I would wash up. Here on this bank, dry out and bleach and the wind would scour and roughen me, strip away the thinnest layers until I was brittle and the thickness of paper. Until I crumbled into sand. That's how I felt. I'd say it was a relief to have at last nothing, nothing, but I was too hollow to register relief, too empty to carry it.
>
> ([2012] 2013, 197)

At other points in the disjointed narrative, Hig finds himself reflecting on where the disease might have come from and what it felt like to anticipate imminent collapse. He recalls:

> I didn't see the collapse, the mass death. But I felt it coming. Like a pressure drop. The kind that is worse than bad weather. We had it a few times growing up at the ranch. A pressure shift you feel in your pulse, your lungs. A darkening of the sky, a weird green tinged blackness. The cattle restless and upset beyond the usual omens of thunderstorm. That's the way it felt. Why I think I should have known.
>
> (252–253)

As in Brooks's *Year of Wonders*, there is speculation in *The Dog Stars* about where the disease might have originated:

> That's what the press reported. Mutation of a superbug, one of the ones they'd been watching for two decades. In the water supply, etc. combined with a bird flu. We called it the Africanized bird flu, after the killer bees. First cases in London and blamed on New Delhi. But that's probably not where it originated. We heard rumors that it originated in Livermore.
>
> (253)

Speculations about where the epidemic began include assumptions that it was somehow a weaponized biological agent, produced with calculation in some foreign country. In reality, thinks the narrator, it probably came from our own weapons facility at Lawrence Livermore National Laboratory in California. "Genetically modifying the flu is an old business", writes Heller (253).

The novel ends with a sense of extreme isolation, a sense of loss and nostalgia. The narrator dreams that his dog, his most steadfast companion, is still alive. And the novel concludes by quoting Li Shang-Yin's 9th-century Chinese poem titled "When Will I Be Home?", which the narrator refers to as his favorite poem. The poem laments,

> When will I be Home? I don't know. [...]
> Someday we will be back together again.
> We will sit in the candlelight by the West window.
> And I will tell you how I remembered you
> Tonight on the stormy mountain.
>
> (Heller [2012] 2013, 319)

As in *Year of Wonders*, there are passages of rhapsodic celebration in *The Dog Stars*, such as when Hig surveys the land below from his small plane: "mountain slopes bunched and wrinkled, wringing themselves into the furrows of couloir and creek, draw and chasm, the low places defining the spurs and ridges and foothills the way creases define the planes of a face" (49). Heller describes this as "the sense of everything in its place", a kind of cosmic order (49).

But by the end of the novel, everything in its place means a kind of cosmic and unalterable disorder or desolation, a profound loneliness that the mind cannot overcome. This emotionally wrenching view of the pandemic (or post-pandemic) experience seems intended as a wake-up call to readers in 2012, a warning of what it will be like when the next pandemic occurs. Although we did not experience anything in 2020–2021 on the scale that Heller imagines, even during the darkest days of mass casualties, I believe many people around the world do feel as if the ground of normalcy has shifted beneath their feet during the time of Covid, and life will never again be the same as before. In the mid-2020s, we are experiencing the sense of vulnerability that Rigby describes in *Dancing with Disaster*, but are we also developing the feeling of multispecies interconnectedness that she says we will need in order to live more mindfully in the future? Some scholars are writing about this (see, for instance, Rigby's "Carnal Relations: Pathogens, Provender, and Embodied Co-presence" [2020]), but I'm not sure this mindset is really taking hold yet.

Finally, let me mention *The End of October* (2020a) by American journalist Lawrence Wright, a political writer for *The New Yorker* magazine. As the author told reviewer Adrian Horton from *The Guardian*, his primary goal was to tell the stories of the "creative, intelligent" scientists to respond to pandemics and similar public health crises with "courage and ingenuity" – to give the backstory of such scientific efforts that is not available to the average person without doing the sort of research Wright often does for his literary and journalistic projects (Wright 2020b, n.p.). This novel tells the story of a fictional Center for Disease Control (CDC) scientist named Henry Parsons, who directs the Center's infectious diseases section. Parsons is sent by the World Health Organization (WHO) to Indonesia to investigate a mysterious disease outbreak in a kind of prison camp outside of Jakarta, and he quickly realizes he has encountered an extremely deadly, highly contagious disease that he does not recognize. While quarantined at the camp after exposure to the Médecins Sans Frontières doctors who have died at the camp, Parsons realizes that the Indonesian taxi driver who brought him to the camp has returned to Jakarta and then flown to Mecca, Saudi Arabia, to participate in the annual Hajj pilgrimage. Parsons follows this man to Mecca, working closely with Saudi health authorities, but they are too late to find Bambang before he begins interacting with his fellow pilgrims. As Parsons tells another doctor in Saudi Arabia when they begin to see cases of hemorrhagic fever:

> Three million people. […] Tomorrow they begin to return to their homes – in Morocco, China, Canada, South America, even on the smallest islands in the Pacific and little villages in the middle of Africa. But they will not travel alone. They will bring this disease with them. And the whole world will be infected, instantly, with no warning, no time to prepare. The experience we are having right now, in this hospital, will be repeated again and again. Even another week or ten days could make a difference for scientists all over the world who are racing for a vaccine or a cure or anything that could diminish the power of the disease. We've got to buy time, that's all we can do.
>
> (Wright 2020a, 105)

This notion of "buying time" to develop an understanding of the new disease and to discover treatment protocols is at the center of the narrative, but it quickly becomes clear that this particular disease, which goes by the name Kongoli, after the place in Indonesia where it was first seen, has already escaped control and is causing massive suffering and death throughout the world. Dr. Parsons, at the behest of the WHO and the CDC, chases the disease from Indonesia to Saudi Arabia, and by the time he realizes it is too late for him to help contain the widening pandemic in Saudi Arabia, he cannot travel home by any conventional means, as all flights have stopped. With the aid of the Saudi Minister of Health, he makes his way to a US naval base in near Bahrain and is able to board a

US submarine that is in need of a medical doctor to help manage the virus among the sailors. He hopes the submarine will eventually bring him back home to the US, but meanwhile he must develop a medical treatment by experimenting with attenuated applications of live virus on the submarine captain's pet birds and, eventually, on himself.

As Dr. Parsons travels the world to inspect and intervene in outbreaks like the Kongoli situation, the narrative also moves back and forth between the story of Parsons's family in Atlanta, Georgia, and various episodes of geopolitical intrigue with US government officials and other world leaders attempting to determine where the disease originated (a common theme in pandemic literature) and whether it might be an act of biowarfare gone awry or merely the result of natural infection. In a conventional adventure novel, one might expect Dr. Parsons eventually to be reunited with his wife and children in Atlanta as the pandemic comes to an end, but in *The End of October*, nothing happens quite so neatly and happily. This also seems to be a tendency in pandemic literature – the subject is so morose and gruesome that there is seldom a simple and happy outcome, as can be seen in the several examples I have provided in this chapter.

We learn suddenly that Parsons's wife, Jill, has died of the disease, as have her sister and brother-in-law. Parsons's two young children – around 10 or 11 years old – are left to take care of themselves in Atlanta, fending off intruders in their house and desperately struggling to find enough food to survive. They have assumed that their father has also died, as he has not returned from his assignment in Indonesia and Saudi Arabia. What comes through with particular vividness in this novel, which is presented in a highly realistic narrative mode, is that pandemics are unpredictable and deadly, they show no mercy, they turn citizens against one another, and the successful response to a pandemic requires as much inter-agency and international cooperation as possible. Although the novel was completed, I believe, prior to the Covid-19 pandemic, the message of cooperation seems to be a commentary against the actual actions and philosophy of the Donald Trump administration in the USA, which actively undermined the efforts of the WHO in response to Covid (by withdrawing US support from the organization).

There is much in Lawrence Wright's *The End of October* that rings true for readers who have lived through the Covid-19 pandemic – the shortages of certain practical commodities (including food), the frustration at the lack of an effective medical response (the slow and inconsistent rolling out of vaccines throughout the world is an example of this), and the suspicion that this tragic public health crisis could be the result of intentional hostile actions by other nations. In the novel, Putin's Russia is presented as the possible source of the Kongoli virus. In the end, though, it seems that Kongoli developed among prehistoric mammoths (giant elephant-like animals) on October Revolution Island in the Arctic Sea, where it in turn infected polar bears, and spread from them to migrating cranes, who brought the disease south where it began spilling over to human populations, the latest in a series of bird flu outbreaks. I am not sure I would place *The End of October* in the same artistic category as highly poetic and deeply psychological works of literature like Brooks's *Year of Wonders* and Heller's *The Dog Stars*, but I do find the more recent novel to be quite a compelling human story and a very interesting way to present information about the history of pandemics and vaccines and about the way public health organizations like the CDC and the WHO go about responding to pandemics. Although Wright recognized that many people might not be "emotionally ready" to read his novel when it appeared during the Covid pandemic, he felt it was important for him to tell a story that recorded the kinds of societal shortcomings we witnessed in the early 2020s: "the partisanship, the needless international antagonisms, the lack of preparation, the disrespect for science, all of those things [that created] havoc we shouldn't have to deal with" (Wright 2020b, n.p.)

Sharing the planet: living mindfully and self-critically with other species

In conclusion, all of these novels reinforce the claims Rigby makes in *Dancing with Disaster* about how pandemic literature might inspire us to live more mindfully and self-critically on a planet that we share with many other species. The purpose of this literature is not to provide specific, practical advice about overcoming infectious diseases, but works like *Year of Wonders* and *The Dog Stars* do reveal certain clues about how to seek out moments of joy even in terrible circumstances. And all three of the novels I have described in this chapter indicate that it is vital for us to appreciate the fact that human beings are not exceptional – that we belong to the animal kingdom and our own fate is interwoven with the fates of our fellow creatures.

American author David Quammen (2012) concludes with a similar message in his powerful book about the interspecies spread of disease, *Spillover: Animal Infections and the Next Human Pandemic:* "That's the salubrious thing about zoonotic diseases", writes Quammen:

> They remind us, as St. Francis did, that we humans are inseparable from the natural world. In fact, there *is* no "natural world," it's a bad and artificial phrase. There is only the world. Humankind is part of that world, as are the ebolaviruses, the influenzas and the HIVs, as are Nipah and Hendra and SARS, as are chimpanzees and bats and palm civets and bar-headed geese, as is the next murderous virus – the one we haven't yet detected.

> I don't say these things about the ineradicability of zoonoses to render you hopeless and depressed. Nor am I trying to be scary for the sake of scariness. The purpose of this book is not to make you more worried. The purpose of this book is to make you more smart.
>
> (2012, 518)

I believe this is the overarching purpose of pandemic literature: to make us smarter, or more mindful of the meaning of our lives in a multispecies and complicated reality. Quammen's more recent pandemic study, *Breathless: The Scientific Race to Defeat a Deadly Virus* (2022), further illuminates not only the multispecies phenomenon of zoonosis but the complex and breathlessly rapid pace of international scientific collaboration – another form of complex global interaction – in response to public health crises. The truth is we are a vulnerable species and we must learn to respect and behave ethically toward the other species (and other human beings) who inhabit the earth with us, as is frequently emphasized in environmental humanities scholarship (see, for instance, Donna Haraway [2016] and Thom van Dooren and Deborah Bird Rose [2017]). In June 2020, *Bifrost Online* published 20 essays on the emerging pandemic, including an "open letter" titled "An Environmental Humanities Response to the Covid-19 Pandemic", which opens with the assertion: "And now a virus carries the message of our interbeing – across bodies, species, continents" (Adamson et al. 2020). This emphasis on relationships, on "interbeing", is one of the core messages of the environmental humanities, even during non-pandemic times, as is particularly emphasized in such subfields as posthumanism and empirical ecocriticism. We must protect the habitat of other species and keep our distance from species that may be harboring dangerous viruses – in many instances, this is something we can do, if we are mindful and careful. We must also think of our fellow humans as compatriots and collaborators, not as rivals and aliens – and in sharing information and resources, rather than merely casting suspicion toward those whom we assume wish to do us harm, we can do a better job of evading and overcoming the future pandemics. Pandemic literature is not intended to offer specific advice about how to solve public health crises, but it can help us to appreciate the salutary psychological lessons of the recent crisis and it has the potential to make us more mindful and self-reflective when responding to the pandemics (or epidemics) we will face in the future.

References

Adamson, Joni et al. 2020. "An Environmental Humanities Response to the COVID-19 Pandemic: An Open Letter." In "The New Normal? An Environmental Humanities Response," edited by Steven Hartman, Joni Adamson, Greta Gaard, and Serpil Oppermann, June 8, special issue. *Bifrost Online*. https://bifrostonline.org/environmental-humanities-response-to-covid-19/

Brooks, Geraldine. 2001. *Year of Wonders*. London: Penguin.
Buell, Lawrence. 2005. *The Future of Environmental Criticism: Environmental Crisis and Literary Imagination*. Oxford: Blackwell.
Camus, Albert. [1947], 1948. *The Plague*. Translated by Stuart Gilbert. New York: Random House.
Clark, Nigel. 2011. *Inhuman Nature: Sociable Life on a Dynamic Planet*. Thousand Oaks, CA: Sage.
Derrida, Jacques. 1984. "No Apocalypse, Not Now (Full Speed Ahead, Seven Missiles, Seven Missives)." *Diacritics* 14 (2): 20–31.
Haraway, Donna. 2016. *Staying with the Trouble: Making Kin in the Chthulucene*. Durham, NC: Duke University Press.
Heller, Peter. [2012], 2013. *The Dog Stars*. New York: Vintage.
McCarthy, Cormac. 2006. *The Road*. New York: Vintage.
McFarland, Sarah E. 2021. *Ecocollapse Fiction and Cultures of Human Extinction*. London: Bloomsbury.
Nayar, Pramod K. 2017. *Bhopal's Ecological Gothic: Disaster, Precarity, and the Biopolitical Uncanny*. Lanham, MD: Lexington.
Nayar, Pramod K. 2019. *Ecoprecarity: Vulnerable Lives in Literature and Culture*. New York: Routledge.
Quammen, David. 2012. *Spillover: Animal Infections and the Next Human Pandemic*. New York: Norton.
Quammen, David. 2022. *Breathless: The Scientific Race to Defeat a Deadly Virus*. New York: Simon & Schuster.
Rigby, Kate. 2015. *Dancing with Disaster: Environmental Histories, Narratives, and Ethics for Perilous Times*. Charlottesville: University of Virginia Press.
Rigby, Kate. 2020. "Carnal Relations: Pathogens, Provender, and Embodied Co-presence." In "The New Normal? An Environmental Humanities Response", edited by Steven Hartman, Joni Adamson, Greta Gaard, and Serpil Oppermann, June 8, special issue. *Bifrost Online*. https://bifrostonline.org/kate-rigby/
Shelley, Mary. [1826] 2004. *The Last Man*. Introduction and notes by Pamela Bickley. Ware: Wordsworth Editions ltd.
Slovic, Scott. 2020. "COVID World, COVID Mind: Toward a New Consciousness." In "The New Normal? An Environmental Humanities Response" , edited by Steven Hartman, Joni Adamson, Greta Gaard, and Serpil Oppermann, June 8, special issue. *Bifrost Online*. https://bifrostonline.org/scott-slovic/
Van Dooren, Thom, and Deborah Bird Rose. 2017. "Lively Ethography: Storying Animist Worlds." In *Environmental Humanities: Voices from the Anthropocene*, edited by Serpil Oppermann and Serenella Iovino, 255–271. London: Rowman & Littlefield International.
Wallace, Molly. 2016. *Risk Criticism: Precautionary Reading in an Age of Environmental Uncertainty*. Ann Arbor, MI: University of Michigan Press.
Wright, Lawrence. 2020a. *The End of October*. New York: Knopf.
Wright, Lawrence. 2020b. "'It's unnerving': Lawrence Wright on the Eerie Prescience of his Pandemic Novel [Adrian Horton." *The Guardian*. May 6. n.p. https://www.theguardian.com/books/2020/may/06/lawrence-wright-end-of-october-interview-pandemic-novel

Part III

Affective ecoprecarity

Relationality, resilience and resistance

8 Of ecological critique and queer utopias

Nicky Drayden's *Escaping Exodus* (2019)

Stefan Benz

Introduction: imagining the precarity of future habitability

Nicky Drayden's (2019) space opera *Escaping Exodus* uses speculative fiction to propose the following thought experiment: what if future humans had to leave a resource-depleted Earth behind and had to survive by colonizing the insides of giant space organisms instead? The novel imagines a matriarchal society that has inverted patriarchal gender roles and that rests upon a strict class hierarchy. Drayden's narration introduces this future human society to readers via the perspectives and the relationship of two young women: Seske, the future leader and matriarch, and Adalla, a member of the working class and Seske's closest friend and romantic interest. Over the course of the story, both protagonists come to experience first-hand the toll the strategies for survival in space are taking on the lower classes of their society and on the "Zenzee" (Drayden 2019, 289), the sentient moon-sized animals they inhabit. Severe classism allows for large parts of the workforce to be disposed of when resources become scarce, and the century-long exploitation of the Zenzee has left this creature's population significantly diminished and close to extinction. The novel, therefore, creates two intertwined coming-of-age stories that trace the development of Seske and Adalla, respectively, from unwitting members of their society into two individuals who decide to work towards radical political, social, and environmental change.

Adalla and Seske thus function as effective seismographs for precarity, that is, for how precariousness or vulnerability is unevenly distributed within their society (Butler 2009, xvii; Schmidt-Haberkamp and Gymnich 2022, 4). Their perspectives reveal that Judith Butler's (2015a) assessment that precarity is "not a passing or episodic condition, but a new form of regulation that distinguishes" the contemporary historical moment (vii) also holds true for this future society. Unwilling to give up their political and social privileges and luxurious lifestyles in the face of constant ecological precariousness, the political elite deprive the larger

DOI: 10.4324/9781003383598-12

parts of the population of "social and economic networks of support" and subject them to "heightened risk[s] of disease, poverty, [… and] violence" (Butler 2009, 25–26; see also Schmidt-Haberkamp and Gymnich 2022, 4). *Escaping Exodus* indeed appears to subscribe to Butler's differentiated notions of precarity and precariousness. The novel not only imagines how the ruling class deflects vulnerability from itself and directs it toward the poor and the nonhuman (2015b, 33), but it also underlines the interdependence of all beings – both human and nonhuman – by foregrounding how all life is equally subjected to precariousness (42).

The political elite's lifestyle also has disastrous consequences for the creatures they colonize. On average, a Zenzee's life force is drained after only 12 years, leaving humans to evacuate and deconstruct – or in the case of the lower classes destroy – their lives and homes, and to relocate to a new host. The novel's plot is triggered by the matriarch's failure to choose a suitable host, for the Zenzee she chooses is pregnant and hence this Zenzee's life force depletes even more rapidly than usual, constituting a severely unstable and precarious environment for her parasitic invaders from the very start. *Escaping Exodus* thus imagines the culmination of a vicious cycle of what Pramod K. Nayar (2019) calls "ecoprecarity" – a term used to focus on both "the precarious lives humans lead in the event of ecological disaster […] and [… on] the environment itself which is rendered precarious due to human intervention in the Anthropocene" (7). This renders the ideological divisions and social precarity within their society drastically clear to Adalla and Seske and feeds their desire to create cosmological and political change. It is, however, not until the two come into close, intimate contact with the space-faring Zenzee, learning first-hand of their sentience, their emotional and cognitive capacities, that they begin to take action.

In a first step, this chapter uses precarity as an analytic tool in order to discuss how *Escaping Exodus* constructs a strong critique of sexism and heteronormativity, classism, and anthropocentrism, as sowing ideological divisions that undergird, legitimize, and exacerbate contemporary Euro-American extractive practices. It is further interested in how Drayden's novel explores intimacy – between homo- and heterosexual partners, members of different classes, between different species – as a means to highlight universal interdependencies, to mark sexism and heteronormativity, classism, and speciesism as unethical, and to work against them. Hence, Steven Shaviro (2021) has fittingly purported that the society portrayed in the novel is not "easily described along a utopian/dystopian axis" (n.p.). By resorting to queer ecological discourse and the work of José Esteban Munoz (2009), more specifically, his notion of queer utopia, as analytical frameworks, this chapter investigates the cosmological and political alternatives *Escaping Exodus* offers.

Of ousted men, imperiled workers, and culled "beasts"

One of *Escaping Exodus's* most distinct speculative achievements is its reversal of traditional Euro-American gender hierarchies. In this matriarchal society, men face inequality on various levels: they are, for instance, not allowed to hold any political office, and the public is segregated along gender lines, requiring them to use separate entries for establishments such as shops (Drayden 2019, 138). Stereotypes about the intellectual, emotional, and physical inferiority of men are rampant (132). Family life is heavily regulated. Each family unit must consist of ten members: nine adults and one child, and the one-child policy is strictly implemented with very few and consequential exceptions. It is the novel's discussion of family life that complements its dystopian vision with a utopian patch (Shaviro 2021, n.p.): while there are rigid rules as to which members of a family unit are allowed to reproduce and with whom they can be in a romantic and sexual relationship, both hetero- and homosexual relationships are the norm. Sexual desire is thus accepted outside of the heteronormative notion of reproduction. This does not mean that sex outside of reproduction is not subjected to functional purposes, as Seske's "ninth sex talk" by the matriarch herself demonstrates: "Sex is about power. About position. About securing a future for our matriline" (Drayden 2019, 75). This view aligns with the classist structures that intersect with and determine marital and sexual life, as Seske and Adalla are made painfully aware of. When their romantic feelings for each other surface, they are faced with enormous push-back and realize that cross-class marriage is considered a "scandal" (61).

Social precarity, the unequal distribution of social precariousness (Butler 2015b, 33), is most clearly demarcated along the lines of class. The political elite, the "Contour class" (Drayden 2019, 88), has successfully created a strict class hierarchy that sustains the luxurious lifestyle of its members. While Seske's perspective provides readers with insight into the privileges and the mindset of the Contour class, Adalla's allows them to learn more about the different lives of workers. Born into an upper working-class family of major organ workers, who prepare and control the Zenzee's major organs, such as the heart, to render human life inside the Zenzee possible, Adalla enjoys comparatively great privilege as one of her mothers explains to her:

> There is a hierarchy among beastworkers, Adalla girl […] we're at the top of that pile […] you have never known hunger. You've never had to live a life of scavenging. Never worried about having to work any of those lesser organs that slowly fill your body with poisons and tumors.
>
> (128–129)

Drayden thus reveals the extremely classist structure of a society that strictly and categorically distributes both financial wealth and risk to the detrimental disadvantage of the lower classes.

The classist-necropolitical ideologies and practices (Mbembe 2003, 11) of this society are most explicitly revealed by the novel's focus on the lowest class, the "grisettes" (Drayden 2019, 85). They are artificially bred as "temporary beastworkers" (171), serving the sole purpose of rendering the Zenzee habitable. Once the process is complete, they are almost clinically disposed of in a ritual classicide, because those in power have agreed that keeping them alive defies economic logic: "there is no room, no resources for them. No work for them to do. It's unsustainable" (172). Their bodies are then rationally recycled as plant fertilizer (172). To justify this treatment, society has systematically segregated them and deprived them of the right to personhood: they are considered "tools, not people" (128), they are forbidden to speak (125), and they are housed in living quarters located in the Zenzee's bowels – a place not visited by members of other classes, for "[w]here else would you put a bunch of people you want to pretend don't exist?" (61). Adalla's quick rise and success as a heart worker is followed by her equally rapid fall when she refuses to participate in classist discrimination against lower-class workers. Relegated to the rank of a boneworker, she acquires knowledge about the fate of the "grisettes" which encourages her to incite political unrest amongst the lower classes that protest the classist oppression by the political elite (197). She uses, in Judith Butler's (2015a) words, "precarity as activism", "rally[ing] precarity against those […] that seek to augment their power to manage and dispose of populations" (xi). Their protest fails at first, but it does lead toward a reconsideration of the fate of the grisettes and of classist discrimination more generally, once Seske assumes the role of matriarch at the end of the novel.

Drayden's triptych critique of Euro-American political, social, and economic practices is completed by the novel's extensive discussion of speciesism and the precarity of the nonhuman. It represents ecoprecarity by reversing the genre template of what Nayar (2019) labels "invasion narrative" (12): it is not a human but a nonhuman body, the Zenzee's, that is "punctured" and "transformed into a host" (12). In order to render the insides of their Zenzee what they consider habitable, humans harvest her internal structures and severely disrupt her metabolism, exploiting, for instance, the Zenzee's bones as building material and triggering a constant fever to create warmer temperatures (Drayden 2019, 153–154, 246). As a result, her centuries-long life-span is cut short to 12 years during which she is subjected to immense pain. Drayden thus allegorizes ecoprecarity to address the vulnerability of the global ecosystem on planet Earth produced by human exploitative practices – ignorant of or indifferent to the disastrous effects these practices have on all lifeforms (Nayar 2019, 14).

Furthermore, *Escaping Exodus* draws attention to how anthropocentrism and human exceptionalism undergird speciesism by upholding an ideological hierarchical separation of the human and the nonhuman: "Never in all my [Adalla's] dizzy dreams had I thought that our beast was something other than a thing, an animate object, a sustainer of life" (Drayden 2019, 78). Only late in the novel are the Zenzee addressed by the name they use for themselves (as is revealed to Seske who learns to communicate with their Zenzee) rather than by the derogatory denomination of "beast" that invokes the dichotomous binary of the civilized (hence morally superior) human versus the primitive (hence morally inferior) nonhuman (12; Wolfe 2003, 17). This is also reflected in Adalla's and Seske's switch from the personal pronoun "it" to "her", signaling that the Zenzee are no longer viewed as exploitable objects but as subjects which are attributed nonhuman personhood and moral standing. Drayden's novel thus adopts a posthumanist mode of thought that Cary Wolfe (2010) claims is not only interested in "the problem of anthropocentrism and speciesism" but also in "how practices of thinking [...] must change in light of their critique" (xix). It is important to note that these processes occur within the minds of Seske and Adalla – a large-scale shift in societal ethics does not yet occur within the novel. Instead, *Escaping Exodus* is invested in imagining how their, and above all Seske's, intimate encounters with the Zenzee trigger what may be termed a queer ecological reconsideration of their cosmological and practical stance toward the nonhuman.

Of ecological intimacy and queer utopias

Timothy Morton's (2010) proposition for a merging of the fields of ecology and queer theory commences with the assessment that both fields of intellectual inquiry "demand [...] intimacies with other beings" (273). To both, intimacy is the remedy for what Morton terms "the inside-outside manifold", which he considers "fundamental for thinking the environment as a metaphysical, closed system – Nature" (274). In Euro-American thought, this manifold has served to suggest clear boundaries between what is male and female as well as what is natural and unnatural. Morton builds on Judith Butler's idea that the body is not such a "closed system" (Morton 2010, 274; Butler 1990, 133–134), and extends it to "all life-forms", who "along with the environments they compose and inhabit, defy boundaries between inside and outside at every level. [...] When the environment becomes intimate [...] it is decisively no longer an environment, since it no longer just happens around us" (Morton 2010, 274). To grapple with such a view on human and nonhuman life-forms terminologically, Morton proposes the notion of liquidity:

> Life-forms are liquid: positing them as separate is like putting a stick in a river and saying, 'This is river stage x' (Quine). Queer ecology requires a vocabulary envisioning this liquid life. I propose that life-forms constitute a *mesh*, a nontotalizable, open-ended concatenation of interrelations that blur and confound boundaries at practically any level: between species, between the living and the nonliving, between organism and environment.
>
> (2010, 275; emphasis in original)

Morton thus contributes, via the idea of liquidity, another image to the corpus of disruptive notions that seek to upend the Euro-American philosophical category of a stable and unified human subject (Schwab 1994, viii–ix) who can live independently of their environment.

Morton portends, however, that "visualizing [this] mesh is difficult: it defies our imaginative capacities and transcends iconography" (2010, 276). It is precisely this challenge of venturing beyond Euro-American iconography that *Escaping Exodus* appears to take up. Its very premise of humanity's colonization of a giant space organism facilitates a creative perspective that emphasizes the impossibility of humans existing separate from their environment. In *Escaping Exodus*, they live because of their environment, and they exist literally inside their environment. The novel thereby aligns with Quine's and Morton's position on the impossibility of conceptualizing the life-forms within an ecosystem by dissecting it, but instead as a "mesh", only chartable as a "nontotalizable, open-ended concatenation of interrelations that blur and confound boundaries at practically any level" (Morton 2010, 275). *Escaping Exodus*, therefore, imagines this figurative river, the life-forms on the Zenzee as a literal river of blood, ichor, and various sorts of the Zenzee's other liquid excretions that threaten and sustain human life inside her. This river becomes Drayden's rhetorical figure of choice to remind readers of the universality of precariousness that determines the life of all beings in the novel and to signal their interdependence (Butler 2015b, 42).

Liquidity thus becomes an important trope for *Escaping Exodus* to signal the enmeshment of the human within the nonhuman. At the beginning of the novel, Seske observes "excavation" (Drayden 2019, 1) – the process during which the Zenzee is rendered habitable – for the first time. While exploring, she is surprised to not (yet) find the comfortable "walkways cobbled with pearlescent stones" she is used to, but to encounter, instead, "a hostile, overgrown marsh" (13) filled with "thick tangle[s] of reeds resting in a swamp of putrid burbling juices" (15). The Zenzee's liquid innards not only appear as uncanny but also as a serious threat to humans, as an accompanying guard reveals to Seske: "this place hasn't been neutralized yet […]. We're basically slowly digesting in this beast's stomach acids." (18) During her training to become a heart worker, Adalla is confronted with another dimension of this threat: working in the arteries, it is crucial to keep track of the Zenzee's heartbeat, for when it hits, blood gushes through these vessels with a force that easily drowns the careless. As one of Adalla's trainers explains, it is quite common for trainees not to survive their first day at work: "One or two [catche] on before the beat c[omes]. The others [enrich] the beast's blood, making the beat stronger for us all" (40). Liquidity thereby further underscores the interrelation between "the living and the nonliving" (Morton 2010, 275), and most poignantly so, in the case of the gruesome fate of the grisettes, who – once no longer needed – are "recycled" by being fed a "cell destabilizing agent" that turns them "into pools of dense fertilizer for the plants" (Drayden 2019, 172) and hence eventually sustenance for other humans.

Drayden thus uses liquidity to blur binaries that pertain to human-nonhuman relationships and to speculatively map the vast and intricate dimensions of human-nonhuman interdependency and interconnectedness. By chance, Seske finds herself in the womb cavity of their Zenzee, where she learns of their Zenzee's pregnancy and is pulled into the womb by its unborn offspring (72). Expecting to suffocate, Seske finds that she can breathe the Zenzee's amniotic fluid that fills the womb, which disrupts her impression of nonhuman/Zenzee liquids as a threat and most tangibly dispels the notion of the (human) body as a "closed system" (Morton 2010, 274; see also Butler 1990, 133–134), impenetrable and self-sustaining. During their encounter, the Zenzee embryo connects and communicates with her through its tendrils. Seske returns to visit the embryo again, and the two enter into a close and evermore intimate relationship, as part of which the unborn Zenzee teaches her to dance the traditional human dances which her family has been unable to (Drayden 2019, 107). The growing intimacy between Seske and the unborn Zenzee is reflected in the language Seske uses to process their interactions: she increasingly resorts to an anthropomorphizing register, and she recognizes the Zenzee as being alive and "sentient" (111, see also 74, 189).

Seske's encounters with the unborn Zenzee and the adult Zenzee's rapid deterioration due to its pregnancy, which shows itself, among other things, in "weeping walks of blood" (152), makes the extent and consequences of human-exploitative behavior blatantly clear to Seske and Adalla. They are now convinced that their parasitic life-style is neither ethical nor sustainable. Symbiosis as an alternative to means of human survival on the Zenzee still seems impossible to them, but once Seske assumes the role of matriarch after her mother's untimely death, she learns of another human society, the Serrata, whose members are able to live in equilibrium with their Zenzee. She immediately seeks them out to learn how they are able to establish and maintain symbiosis (236). While away, Seske is betrayed by her illegitimately born sister who devises a plot to gain the throne, leaving Seske without a shuttle to return to their Zenzee. Seske is, however, able to reach their Zenzee in time to stop the coup by covering herself with a "honey-covered gel" produced by the Serrata's healthy Zenzee, a gel that protects the Zenzee – and humans – from the cold in space and that is not found on their own, sick Zenzee. In this passage, the novel imagines intimate human-nonhuman contact via liquidity not only to indicate interconnectedness, but also to mark a specific form of human conduct as ethically desirable and mutually beneficial. Once returned, Seske directs all her efforts toward ensuring their Zenzee recovers a healthy status.

Seske's intimate relationship with the Zenzee finally culminates in what Annie Sprinkle and Beth Stephens (2021) call an "ecosexual" act (1). As part of her efforts to save her people and Zenzee from pending destruction, Seske attempts to communicate with the Zenzee by entering an "orifice" close to its liver (Drayden 2019, 278). Once inside, a biochemical substance puts Seske in a dreamlike state in which she experiences being a Zenzee herself and encountering another Zenzee in space, a meeting that climaxes in an act of mating, during which Seske experiences sexual gratification and an "ovispore" – a cell of Zenzee semen – "disappears inside" her (278). She soon finds out that she really has been impregnated by the Zenzee. This realization comes at a significant, intimate moment Seske shares with Adalla, amidst and despite the chaos that defines the events at the end of the novel: "our bodies have found comfortable places against each other, and we just ... fit. [...] No matter what scars the future may hold, I [Seske] want this moment" (285). Yet, the two are interrupted by something they are not expecting:

> 'Seske,' [Adalla] says. 'I think something weird is happening.'
> 'Nothing weird is happening.'
> 'You're wet. Down there.'
> 'I think that's pretty normal.'
> 'Like *really* wet.'
> 'And that's good, right?'
> 'Seske?'
> 'Yes, Adalla?'
> 'There's some kind of tentacle thing curling around my thigh.'
> (286; emphasis in original)

In the next moment and interrupting the first sexual encounter between them, a "dozen black tentacles erupt from [Seske's] humble bits" (287).

She remains unexpectedly calm, thanks to the Zenzee's hormones that are at work inside her body (294). Instead of panicking, she suddenly appears to know of a way to redeem her society's long-lasting exploitative behavior toward the Zenzee: she needs to ensure the safe transport of their Zenzee's semen to another's womb.

The novel thus interweaves the occurrence of intimacy between humans and between the human and the nonhuman to identify a utopian moment of queer transgressive possibility – of cross-class romance and of interspecies fertilization. It defies the idea that reproduction and survival depend entirely on intraspecies heterosexual intercourse – as has been proven by biological studies (Morton 2010, 276). Instead, it follows Morton's conclusion that sexual practices are "liquid", transcending species boundaries, and proving that "evolution is a story of diverse life-forms cooperating with one another" (276). Notably, the Zenzee plays an active part in this "ecosexual" encounter (Sprinkle and Stephens 2021, 1), displaying agency by impacting Seske's thoughts and actions through the hormones that she implanted within her. The novel thereby gestures toward symbiosis as a viable model for human-nonhuman cohabitation, by making use of queerness as "a reserve of utopian energy" (Morton 2010, 280) for cosmological, social, and environmental regeneration.

The relationship between the notions of queerness and utopia that *Escaping Exodus* develops may be best understood via the work of José Esteban Munoz. In his 2009 study *Cruising Utopia – The Then and There of Queer Futurity*, Munoz endeavors to reconceptualize the category of utopia within the framework of queer studies. His definition of queerness features two components: first, it is understood as being "essentially about the rejection of a here and now" (1), because it is the here and now, the present (Euro-American) "version of reality that naturalizes cultural logics such as capitalism and heteronormativity" (12) and rejects queerness. Therefore, "queerness [remains] an ideality" (1). Munoz does not deny that queerness occurs within this queerphobic (Euro-American) present, but he claims that it can only exist within brief and fleeting utopian moments. Hence, queerness may be understood as "that thing that lets us feel that this world is not enough, that indeed something is missing" (1). To him, art is an important medium that can create and make tangible queerness as something that is not yet here:

> Often we can glimpse the worlds proposed and promised by queerness in the realm of the aesthetic. The aesthetic, especially the queer aesthetic, frequently contains blueprints and schemata of a forward-dawning futurity […] queer aesthetics map future social relations.
> (Munoz 2009, 1)

Second, queerness represents "an insistence on potentiality or concrete possibility for another world" (1). With Ernst Bloch, Munoz develops the category of "concrete utopias" which are always "relational to historically situated struggles" (3). Concrete utopias manifest as "hopes of a collective", they can "be daydream-like" but they can also be "dream[t] for the many" by a "solitary oddball" (3). To better recognize and read such concrete utopias, Munoz proposes to rely on "hope as a hermeneutic" (4).

Aside from its critique of sexism and heteronormativity, classism, and anthropocentrism, *Escaping Exodus* is majorly invested in imagining queerness "as a rejection of [this] here and now" and as a schema "of a forward-dawning futurity" (Munoz 2009, 1). Queerness, as it is represented by intimacy between humans and between humans and nonhumans, allows Drayden to "map future social relations" and to make her readers "feel that this world is not enough, that indeed something is missing" (1), which, according to the novel, is humanity's striving for and implementation of symbiotic relationships with their nonhuman environment. *Escaping Exodus* speculatively negotiates how such concrete utopias can become reality. Both Adalla and Seske are indeed "solitary oddballs" (3) – both are unwilling to fulfill society's expectations – who share the hope for significant political and social change, who do not only "dream for the many" (3), but whose dreams inspire and eventually incite these changes in the real.

Above all, it is Adalla's and Seske's romantic relationship that marks the novel's vision as queer and as inviting the readers' reliance on "hope as a hermeneutic" (Munoz 2009, 4). Right before they attempt to engage in their first sexual encounter, Seske proclaims: "What I am now is just hopeful. Mostly hoping that I can kiss you [Adalla] without something weird happening" (Drayden 2019, 286).[1] Their initial hope of being together is at first youthfully optimistic, but soon disappointed when both are forced to go their separate ways. For most of the novel, it hangs by a thin thread, kept alive only by their unlikely episodic meetings and memories of their time together. As Munoz would say, they successfully equip themselves with "the past [memory] and the future [hope] as armaments to combat the devastating logic of the world of the here and now" (2009, 12). In the end, this allows them to find each other again, even though Adalla, as the leader of a workers' rebellion, and Seske, as the new matriarch, are by then the heads of opposite political sides. It is their love that inspires them to work together toward stopping the Zenzee's decline and setting their society on the course for symbiosis. For its discussion of queerness, utopia, and hope, *Escaping Exodus* therefore fittingly employs Adalla and Seske as homodiegetic narrators, in order to transmit their sensibilities most immediately and effectively.

Many of the events in *Escaping Exodus* are of epic significance, but at the same time, and like the art in which Munoz's *Cruising Utopia* is interested, the novel draws attention to the ideas that queerness is intricately connected to sexual desire and that "utopia exists in the quotidian" (Munoz 2009, 9). After Adalla learns of and personally witnesses the classicide of the grisettes, her faith in the system that she grew up in is broken for good. She is unable to cope with her pain, drowning it unsuccessfully in alcohol and other intoxicants. The only person able to console her is Laisze, a fellow boneworker. When a particularly heavy fit of nightly desperation hits her, Adalla slips into Laisze's bed to share her grief. During their conversation, Adalla reaches the conclusion that she can no longer go idle in the face of the inequality and precarity she has witnessed: "All throughout our history, we sing two kinds of women ... those born into power and those who disrupt power. I intend on being the latter" (Drayden 2019, 178). Deprived of the right to private quarters, boneworkers sleep in large bunk bed constructions, which make it easy for the guards who are tasked with patrolling and closely surveilling them to eavesdrop. To conceal their conversation, Laisze instructs Adalla as follows: "She throws the covers over our heads and says, 'Moan' [...] I take a moment to gather myself, then let out a smooth, throaty purr" (178).

It is the quotidian (acoustic pretense of) sexual intercourse that allows them to lead their debate on the possibilities of inciting a political revolt to "map [...] utopia" (Munoz 2009, 1) in secret. Hesitant at first, Laisze soon agrees to support Adalla and to make use of her social connections and extensive knowledge about the secretive spaces on the Zenzee to establish a network of rebellion. Significantly, the pretended sexual intimacy between them is not only a successful cover of their plans but a lubricant: "so begin our plans, now pressed up together in [...] bed [...] whispers passing between us like those of doting lovers" (Drayden 2019, 179). As their conversation begins to focus on concrete plans, they resort to sign language to further minimize the risk of being overheard. Thereby, their sexual intimacy becomes less a pretense and more of a reality:

> we are pressed together so close that our signs become more like caresses [...] next thing I know, Laisze's telling me the location of a secret meeting spot, her mouth hot upon mine, her tongue twisting, twirling in my mouth.
>
> (179)

Drayden thus marks queer intimacy as conducive not only to environmental but also to socio-political change toward a concrete utopia.

Escaping Exodus ends not with a prospective look into the utopia that Adalla's and Seske's efforts have engendered, but with another brief and intimate scene between them. Seske announces her sexual advance by returning to their earlier conversation:

> 'I'm going to kiss you and something weird is going to happen. But a nice weird. No tentacles, I promise.' […] I pull her close, and our lips touch, and an instant later, the world stops spinning – no artificial gravity, the last of my demands to Doka. At least while our Zenzee heals. And just like that, we're drifting up, up. Adalla pulls me closer […]. Like all there is of import in this life is within this space heated by our breath.
>
> (Drayden 2019, 300)

In this way, the novel's final scene keeps its readers locked in suspense, performatively confirming that its main project is to foreground the ways in which art can tie together queerness and utopia in order to imagine ways of defying and rejecting the "here and now" and to make visible a "forward-dawning futurity" (Munoz 2009, 1). The project of imagining how Adalla's and Seske's efforts bear fruit Drayden postpones to the second novel in the series, *Escaping Exodus: Symbiosis* (2021).

Conclusion: escaping exodus toward queernatureculture

This brings us to the epigraph from the Christian theologian George Morrison:

> Ecological coexistence is 'nearer than breathing, closer than hands and feet' […]. We have others – rather, others have us – literally under our skin (Clark). This is about symbiosis, but it's also about what Donna Haraway calls 'companion species.'
>
> (Morton 2010, 277)

Intimacy, both between humans and between the human and the nonhuman, drives the rich speculative vision of Nicky Drayden's *Escaping Exodus* and allows it to formulate both a critique and – more importantly – a vision beyond the divisive and exploitative ideologies and practices that have spawned contemporary ecological and social crises. It offers insight into how the human and the nonhuman are closely imbricated in a liquid, but also gaseous, and solid mesh that needs ontological, epistemological, and ethical attention and reconsideration. Steven Shaviro (2021) has praised the novel for "suggest[ing] […] an ontology that is very different from anything we are accustomed to" – an ontology that engenders a productive kind of "cognitive estrangement" (n.p.). In fact, he concludes that the novel is proof that "even our most advanced theoretical articulations have yet to keep pace with the constructions of speculative fiction" (n.p.). Concepts like Donna Haraway's (2003) "naturecultures" or Karen Barad's (2007) "intra-activity," Shaviro claims, may come close, but he finds them to not "seem quite right either" to adequately analyze the novel's toppling of "the traditional binary between [the] biological and [the] social" (n.p.).

In light of the focus *Escaping Exodus* takes on intimacy and sexuality, David Bell's (2010) term "queernaturecultures" (134) may present a more adequate choice for terminologically comprehending the ontology, the epistemology, and the ethics Drayden's novel creates. It quite graphically addresses and responds to the question of whether sex is "nature or culture" (142) with the affirmative "both!" and the imperative not to "devalu[e]" either "sex [or] nature" (143). *Escaping Exodus* may at times fail to consequently disrupt and think beyond binary notions of gender and sexuality; in its brief discussion of trans-identities and sex-change, for instance, it presents these only as either grounded in the motive to further one's political career or as punishment. Still, the novel's focus on queer human and nonhuman intimacies allows it to produce a generative kind of "cognitive estrangement" within readers (Shaviro 2021, n.p.) that pulls them in close and offers them the imaginative capacities for a new iconography that makes a queer ecological perspective comprehensible. In so doing, Drayden's writing confirms the important role of speculative fiction for imagining and (re)negotiating the ways in which humans can live and the ways in which they may want to live in times of severe social and environmental crises and the grossing precarity these crises produce.

Note

1 It is important to note that while their homosexual relationship is not threatened by heteronormativity, it is fair to assume that many readers perceive it as queer due to their own social context which may be subject to heteronormativity.

References

Barad, Karen. 2007. *Meeting the Universe Halfway: Quantum Physics and the Entanglement of Matter and Meaning*. Durham, NC: Duke University Press.

Bell, David. 2010. "Queernaturecultures." In *Queer Ecologies: Sex, Nature, Politics, Desire*, edited by Catriona Mortimer-Sandilands and Bruce Erickson, 134–146. Bloomington, IN: Indiana University Press.

Butler, Judith. 1990. *Gender Trouble: Feminism and the Subversion of Identity*. New York: Routledge.

Butler, Judith. 2009. *Frames of War: When Is Life Grievable?* London: Verso.

Butler, Judith. 2015a. "Foreword." In *State of Insecurity: Government of the Precarious*, edited by Isabell Lorey, vii–xi. London: Verso.
Butler, Judith. 2015b. *Notes Toward a Performative Theory of Assembly*. Cambridge, MA: Harvard University Press.
Clark, J. Michael. 1994. "Sex, Earth, and Death in Gay Theology." *UnderCurrents: Journal of Critical Environmental Studies* 6: 34–39.
Drayden, Nicky. 2019. *Escaping Exodus*. New York: HarperCollins.
Drayden, Nicky. 2021. *Escaping Exodus: Symbiosis*. New York: HarperCollins.
Haraway, Donna J. 2003. *The Companion Species Manifesto: Dogs, People, and Significant Otherness*. Chicago, IL: Prickly Paradigm Press.
Mbembe, Achille. 2003. "Necropolitics." Translated by Libby Meintjes. *Public Culture* 15 (1): 11–40.
Morrison, George H. [1994]. "The Reawakening of Mysticism." In *The Weaving of Glory*, edited by George H. Morrison, 103–111. Grand Rapids, MI: Kregel.
Morton, Timothy. 2010. "Guest Column: Queer Ecology." *PMLA* 125 (2): 273–282.
Munoz, José Esteban. 2009. *Cruising Utopia, The Then and There of Queer Futurity*. New York: New York University Press.
Nayar, Pramod K. 2019. *Ecoprecarity: Vulnerable Lives in Literature and Culture*. London: Routledge.
Quine, Willard Van Orman. 1950. "Identity, Ostension, and Hypostasis." *Journal of Philosophy* 47 (22): 621–633.
Schmidt-Haberkamp, Barbara, and Marion Gymnich. 2022. "Representing Poverty and Precarity in a Postcolonial World – An Introduction." In *Representing Poverty and Precarity in a Postcolonial World*, edited by Barbara Schmidt-Haberkamp, Marion Gymnich, and Klaus P. Schneider, 1–18. Leiden: Brill.
Schwab, Gabriele. 1994. *Subjects without Selves: Transitional Texts in Modern Fiction*. Cambridge, MA: Harvard University Press.
Shaviro, Steven. 2021. "Nicky Drayden, Escaping Exodus and Escaping Exodus: Symbiosis." *The Pinocchio Theory*, February 27. http://www.shaviro.com/Blog/?p=1737
Sprinkle, Annie, and Beth Stephens. 2021. *Assuming the Ecosexual Position: The Earth as Lover*. Minneapolis, MN: University of Minnesota Press.
Wolfe, Cary. 2003. *Animal Rites: American Culture, the Discourse of Species, and Posthumanist Theory*. Chicago, IL: University of Chicago Press.
Wolfe, Cary. 2010. *What Is Posthumanism?* Minneapolis, MN: University of Minnesota Press.

9 Women's writing as eco-translation

The critical-creative edges of precarious presence in Bernardine Evaristo's *Girl, Woman, Other* (2020) and Sharon Dodua Otoo's *Ada's Room* ([2020] 2023)

Sonja Frenzel

Then and now: women writing upon critical-creative edges

"We think back through our mothers, if we are women", asserts Virginia Woolf in her seminal 1928 essay *A Room of One's Own* ([1928] 1945, 88). And indeed, the expanding anglophone corpus of women's writing offers sound foundations to build upon when "we", as women writers, set out to contemplate humankind's precarious presence on earth and to imagine the possibilities for future habitability. And yet, how do "we", as women, think forward? How do we tell stories that appreciate humankind's past and our share in it, while at the same time envisioning new, more holistic and sustainable, more-than human futures? This contribution probes possible answers by delineating a literary tradosphere (i.e. a literary, planetary translation system) that emerges in the precarious spaces between tradition and innovation, between thinking backward and imagining forward planetary habitability, between writerly moments of being and readerly immersion into the stories of the place that we live (see Bennett).

Literature has always precariously balanced on this edge between critically assessing the grand narratives of its own canon and of human presence on earth, while creatively experimenting with new ideas and generating new stories. Hence, when Woolf tackles the multi-faceted theme of "women and fiction", she addresses a historical predicament with deep contemporary resonances. Then and now, the critical-creative edges of literary tradition and innovation foster particular experiences of precariousness for writers (see Eliot 2009, n.p.) and for women writers in particular.

This contribution follows a three-fold aim: firstly, two contemporary novels, Bernardine Evaristo's (2020) *Girl, Woman, Other* and Sharon Dodua Otoo's ([2020] 2023) *Ada's Room*, will be integrated into a literary

DOI: 10.4324/9781003383598-13

tradition of women's writing so as to, secondly, explore how they invest the trope of literary motherhood to articulate temporal and spatial-ecological precarity. Indeed, both novels feature stories of precarity, which nonetheless offer hopeful examples of decolonizing literature, be it in terms of race, gender, or anthropocentrism. Finally, and thirdly, this contribution ties its various argumentative strands together into a reading of literature as more-than-human tradosphere in which literary tradition may co-emerge with writerly and readerly imagination to build a future planetary habitability.

In historical perspective, Woolf laments the lack of a tradition of female authorship in the male-dominated canon of 19th- and early 20th-century literature. The canonical works of "great men writers", she elaborates, may offer women writers "a few tricks" ([1928] 1945, 88); yet, neither their prevalent style nor their generic form readily lend themselves to articulating a woman's experiences of being in the world, nor do they propose suitable models for developing her writerly expression. Rather than aspiring to the "direct" and "straightforward" prose of a man writer (115), women's writing must define its own premises and thresholds. To this end, she advises her fellow women writers to rediscover and to critically acknowledge a lineage of writerly mother figures, such as the Brontë sisters, Jane Austen, and George Eliot, whose exceptional creative achievements may offer guidance for modernist women writers.

Initially, Woolf's argument is thus built upon a notion of gendered privilege that anticipates contemporary debates and foreshadows their inherent controversies surrounding today's identity politics. For women, being a writer has never been a comfortably private or personal affair. Instead, their writerly practice has been – and still is – circumscribed by stereotypes of womanhood: "killing the angel in the house [...] [has been] part of the profession of a woman writer" (Woolf [1942] 2012, n.p.). In order to write, a woman is compelled to dismantle the gendered politics that impinge upon her practice. A canon of women's writing, as Elaine Showalter ([1972] 1992) further explicates from her second-wave feminist vantage, derives from struggling against these restraints: "the authentic line from which women writers trace their descent is one of protest, innovation, and confrontation" (209).[1]

Upon the critical-creative edge of the private and the political, Woolf's writerly oeuvre certainly illustrates the politics of women's lived experience. Her renderings of the stream of consciousness technique thoroughly immerse readers into the female protagonists' minds: Clarissa Dalloway as the perfect hostess in *Mrs Dalloway* ([1925] 1992a), as much as Mrs. Ramsey as mother and wife in *To the Lighthouse* ([1927] 2019), offer nuanced portraits of the above-cited "angel in the house", while the experimental novels *Orlando* ([1928] 1993) and *The Waves* ([1931] 1992b) more ostensibly undermine gendered clichés. All the while, Woolf's writerly practice remains oriented towards the privacy of sublime "moments of being" ([1942]1985, 70). In this way, she adapts the tradition of a seemingly ubiquitous modernist mode of writerly being into her unique experimental style. Woolf's moments of being resemble James Joyce's epiphanies (see Schulkind 1985, 19). At the same time, they evoke a strong sense of the writer's embodied presence, which differs from the spiritual enlightenment Joyce sought. Whether or not this difference originates in the writers' gender is a question that will need to remain uncontested in this chapter. And yet, "amazingly, Virginia Woolf and James Joyce lived identical life spans: 1882–1941. Yet how much he could think and say which she was forbidden!" (Showalter 1992, 219).

Woolf's writing translates the writer's personal and embodied experience into narrative examples of the social and political circumstances of being a woman. Today, the experience of these precarious imbalances continues to resonate profoundly in women's writership. As Woolf has become a literary mother figure for women writing in her wake, her extensive oeuvre delineates the historical grounds for this chapter's explorations into the critical-creative edges as articulated in two instances of contemporary women's writing in English (Evaristo) and German (Otoo). Indeed, as similar arguments on privilege need to be made for writers-of-colour and for writers of the LGBTQ+ community (to cite but two prevalent lines of further diversification), the following explorations will trace the sense of writerly precariousness that has shaped the works of two women writers-of-colour: Bernardine Evaristo, winner of the Booker Prize (2019), and Sharon Dodua Otoo, winner of the prestigious Ingeborg-Bachmann-Preis (2016).

Both women have commented on the critical-creative inspirations as well as on the private-political conundrums that have guided their writerly becoming. In her memoir, *Manifesto: On Never Giving Up* (2021), Evaristo spells out the critical-creative edges of her search for literary mother figures:

> I've noticed that many British writers claim Virginia Woolf, Jane Austen, Emily Dickinson or Henry James as literary influences, yet as a very young woman I felt that I was writing against them. […] Black women writers were the ones I needed to read as a very young woman, and I didn't come across any in Britain who were born or raised here and writing our stories from this perspective. My inspiration came from African Americans: Audre Lorde, Toni Morrison, Gloria Naylor and Alice Walker were foremost among them, and of course Ntozake Shange; and the Jamaican-American writer Michelle Cliff, and the Nigerian novelist Buchi Emecheta.
>
> (157–158)

Almost 100 years after *A Room of One's Own* was published, Evaristo may obviously draw on a more expansive line of literary mother figures. From a diachronic perspective, women's writing has produced its own canon, which, from a synchronic perspective, also recognizes the planetary range of women's writing that still constitutes a blind spot in Woolf's essay.

Anglo-German writer-of-colour, Sharon Dodua Otoo, further expands these premises, when she acknowledges literary mother figures from both the anglophone and the German canon. Her appreciation of writerly mother figures in her acceptance speech for the prestigious Ingeborg-Bachmann-Preis, *Dürfen Schwarze Blumen malen? Klagenfurter Rede zur Literatur* (May Black People Draw Flowers? Klagenfurt Talk on Literature) (2020), not only pays homage to a planetary community of women writers-of-colour; it also diversifies literary historiography, literary tradition, and literary practices.[2] In thinking back through their literary mothers (and the occasional fathers), both Evaristo and Otoo may draw on a critical appreciation of writerly tradition, while they have developed their own unique forms of creative expression. Positioned upon a critical-creative edge, both writers adapt the modernist stream-of-consciousness technique and Woolf's epiphanic moments of being into their novels' articulations of contemporary precariousness and their imaginings of future habitability.

In fact, contemporary writing in general and women's writing in particular poses the inevitable question, "how do we want to live?" This question has often been tackled with recourse to the stereotypical translation from writing and creation to womanhood and motherhood (see most notably, Stanford Friedman 1987). While these observations certainly foster multiple contestations, they also evoke the trope of Earth Mother Gaia as a generative and nourishing source of life, and have thus been taken up in ecofeminism (see Gaard 1993; Levy 2013; Plumwood 1993). Extending motherhood to this planetary level may offer pertinent lines of thinking backward and forward through Gaia. After all, in the light of an ever-accelerating environmental emergency, the future habitability of the earth as the only home our species has ever known, has become – and is becoming – increasingly precarious. By reconnecting our stories with ecologies, writing may regain its meaning as a creative and critical resource positioned upon a pivotal political edge. After all, it may equally challenge us into awareness not only of ecological precariousness and our complicity in it, but also of creative responsibilities.

These responsibilities resonate in *Ada's Room* and in *Girl, Woman, Other*. In Otoo's recent novel *Ada's Room* ([2020] 2023), Ada lives through three double loops of incarnation,[3] in which she experiences epiphanic moments of being and acquires a deeper knowledge of the precariousness of her own life, first as daughter and later as mother. Throughout her spiritual journey, Ada is accompanied and guided by a sentient spirit, who embodies various vibrant and agentic objects, including a bracelet and a broom, as well as material gatekeepers like a doorknob, a room, and a passport. This more-than-human presence translates the narrative beyond human storytelling, as has been foreshadowed in Otoo's prize-winning text *Herr Göttrup setzt sich hin* (Mr Göttrup sits down) (2022), in which an egg tells the story of "Herr Göttrup", sitting at the breakfast table and about to eat it.

In appreciating Otoo's work as existing upon a synchronic critical-creative edge, it is possible to read this innovative writerly practice as one that at once reverberates with Jane Bennett's theoretical concept of "vibrant matter" (2010) as well as with Serenella Iovino and Serpil Oppermann's "storied matter" (2014); besides, it may also find its literary – sisterly – parallel in Ghanaian-American writer Yaa Gyasi's debut novel *Homegoing* ([2016] 2017), in which two halves of a stone connect the family lines of half-sisters Esi and Effia from Africa to the Americas to Europe and back.

Evaristo, in turn, coined the term "fusion fiction" for her characteristic writing in-between prose and poetic genres (2021, 142). In her most recent of eight novels, *Girl, Woman, Other* (2020), her style culminates in an experimental adaptation of the modernist stream of consciousness technique, which "allow[s] [her] to flow freely – from interiority to exteriority, from the past to the present, from one character's narrative to the next" (2021, 142). Furthermore, by thoroughly interweaving the life stories of 11 female and one non-binary character-of-colour, she makes precariousness merge into a shared human experience, yet with diverse articulations, and into planetary entanglements, yet with distinct local groundings. Among the five mother-daughter relationships, the stories of Bummi and Carole exemplify the experience of precariousness in ecological terms: they pivot on the tensions between a neo-liberal economy and environmental exploitation, cultural tradition and global entanglements, petrol-industry and a New Green Economy.

In appreciating this literary work written from a synchronic critical-creative edge, it is notable that Evaristo's most immediate literary inspiration originates in her own theatrical work on African American writer Ntozake Shange's ([1975] 2021) "celebrated 'choreo-poem' *for coloured girls who have considered suicide / when the rainbow is enuf*" (105; emphasis in original), which offers a crucial – sisterly – source of inspiration for developing new literary forms for articulating the diversity of women-of-colour's experiences in Britain and across the globe.

Following in the wake of Woolf and other literary mother figures, Evaristo's and Otoo's writing thrives on the concomitant practices of critically assessing stories of the past and creatively imagining the possibilities of future habitability. In doing so, the novels *Girl, Woman, Other* and *Ada's Room* translate this critical-creative edge from the writers' engagement with the literary canon to the characters' stories as articulated in the literary text: experiences of precariousness feature prominently both in the fictional world-building and in the characters' stories. Moreover, in both narratives, motherhood offers the most pertinent literary trope of precariousness. Neither of them, however, presents motherhood as uncontested. Rather, motherhood, too, balances on a precarious edge, always translating from past experience into future potentiality, as the following discussion will show.

Critical-creative edges of literary motherhood

As a literary trope, motherhood exemplifies the precariousness of critique and creativity: a mother typically embodies the past and holds a family's storied memory, while at the same time she gives birth to children who will propel these stories forward into an unknown and unknowable future. On the one hand, the experience of motherhood itself – as well as the subsequent literary readings of motherhood – offers the possibility to critically assess the known and to creatively imagine the yet unknown. And yet, on the other hand, motherhood has never been uncontested. As Siri Hustvedt ([2020] 2021), points out in her illuminating essay "A Walk with My Mother", motherhood has always been surrounded by often rigid cultural norms and societal expectations:

> The legacy of the Angel of the House in contemporary Western culture lingers in the bland lists aimed at middle- and upper-middle-class women, touted as 'easy' and 'simple', but which have a decidedly punitive undertow – *good* carries the dread of *bad* with it, after all. It is as if there is a critical juncture when one turns into the other.
>
> (36; emphasis in original)

Motherhood, above all, is compelled to maintain a precarious balance between self and other, between individual and community, between past and future. In Otoo's *Ada's Room* and Evaristo's *Girl, Woman, Other*, these critical-creative edges materialize in literary temporalities and literary ecologies, respectively.

For one, Ada's double loops of incarnation in the mid-15th, mid-19th, and mid-20th century open up the long historical view that "all beings – past, present, and future – are connected to one another. That we always were and always will be" (Otoo [2020] 2023, 130). This diachronic connectedness is translated to a synchronic plane in the lives of Nigerian-born Bummi and her London-born daughter Carole. Their stories will be singled out from the five mother-daughter relationships featured in Evaristo's *Girl, Woman, Other* because they not only tackle pressing ecological issues, but stress their planetary scale through the lasting bonds between the former colonies and former "mother city" of the British Empire.

Ada's experience of temporal precariousness exemplifies thinking backward through eye-opening epiphanic moments, in which she gains insight into her own experience of being a daughter and into her respective mothers' courage. Bummi's experience of ecological precariousness, in turn, stresses her ambition to think forward and to discover possibilities for future habitability on a planetary scale.

Temporal precariousness

Ada's Room is not a strictly linear historical novel, but unfolds along three double orbits that are represented in the novel through the symbol for "eternity": ∞. At the crossing of these orbits, "Between the Orbits", Ada finds herself in her fourth incarnation as a single, pregnant woman-of-colour looking for a bedsit in contemporary Berlin. Structurally, this contemporary episode is placed at a turning point in the middle of the narrative that also marks the turning point in the double loops of incarnation: Ada has so far gone through the first set of three incarnations, each of which had been characterized by intense pain and suffering. In all of these previous lives and in her contemporary incarnation, Ada grows up motherless, and although she becomes pregnant in each life, her sons do not survive. In the fourth, contemporary incarnation, too, Ada is pregnant, though this time with a daughter. The precariousness of this experience resonates forcefully in lingering prejudices against single mothers(-of-colour), but also in the utterly overwhelming experience of becoming a mother when she has never known her own (Otoo [2020] 2023, 140).

As her birthmother died when Ada was still a small child, the role of "mothering" was initially taken over by her father's "sisters" in Ghana. This community of elderly women and extended family finds a first parallel in Berlin, where two elderly women-of-colour offer Ada accommodation in their shared flat. Their room and their laughter remind Ada of her "aunts" in Accra (252). Besides, by a curious twist of fate, these women once knew Ada's mother and shed light upon her untimely death: Ijemma died in an arson attack (261).

This revelation initiates Ada's second loop of incarnation. Realizing that there is more to learn about her mothers' deaths and subsequent absences from Ada's life, she returns to each of her previous incarnations to catch up with what she has yet to learn: for one, she bears witness, as a daughter, to her respective mothers' acts of kindness and bravery in which they sacrifice their lives so that Ada may live; furthermore, each of these epiphanic moments of being ends by naming Ada's mothers and by thus recognizing the precariousness of their motherly and womanly presences.

In fact, there are striking historical parallels between her mothers' lives. Most notably, in the mid-15th century, Ada's mother Farida hides Ada and her brother when their village is raided (Otoo [2020] 2023, 115). The children are spared because of their mother's foresight and decision to abandon them, and Ada eventually finds shelter with "the toothless", a community of elderly women by the Ghanaian coast – offering a second parallel to her contemporary immersion into a community of women-of-colour.

The double loops of incarnation in Otoo's *Ada's Room* blur the distinctions between past and present. In its multiple translations across time, Ada's story continues the story of her mothers and her previous selves, while she herself, as a pregnant mother, will also find her life continued in her yet unborn daughter. In biographical and in collective perspective, therefore, her life is integrated into a shared experience of womanhood and motherhood alike. Both "the child and the old woman belong to the single trajectory of a life in time, early and late" (Hustvedt [2020] 2021, 33). It is noteworthy, though, that this connectedness does not depend on motherly love, but may prosper fully in recognition of a wider – communal – notion of mothering.

Indeed, as Ada's spiritual companion summarizes in response to Ada's sense of being overwhelmed by becoming a mother: "Did she really think that only the person who had born her could be a mother to her? After all her lives?" (Otoo [2020] 2023, 140). As the idea of motherhood continuously expands from biological motherhood to a larger community, Ada's soul companion with its multiple personae themselves become recognizable as a mothering presence. Together with an omnipresent, female God, the personae guide Ada through her incarnations without predetermining her path. Ada's incarnations and their contemporary culmination point thus undermine not only the temporalities of our human presence, but they also subvert prevalent Western-European conceptualizations of being human. Along with Ada, the reader is compelled to realize that "[e]very human being is a complex amalgam of features we recognize only because they are repeated in time" (Hustvedt [2020] 2021, 33). Human presence, therefore, is entangled into an ecology of human and more-than-human presences, each infinitely continuing the other.

Such continuities may also be detected in the mother-daughter relationships of *Girl, Woman, Other*. In her old age, when she has found success and happiness in London, Bummi "wishes her mother was alive to enjoy her new life / see me now, Mama, see me now" (Evaristo 2020, 188). Her daughter Carole, in turn, has inherited both her mother's and grandmother's resourcefulness as well as their ambition to think forward. Besides this diachronic connectedness, the relationship between Bummi and Carole exemplifies the two women's synchronic entanglements with planetary ecologies of Earth Mother Gaia.

Ecological precariousness

The trope of Earth Mother Gaia has typically been associated with images of creation and nourishing resourcefulness. In the light of an ever-accelerating and human-made environmental emergency, however, it needs to be recognized, too, as an emblem of suffering and pain. The future habitability of our planet has become utterly precarious. Thinking back through Gaia compels humankind to (re-)acknowledge the earth as life-sustaining habitat, while thinking forward with Gaia challenges us to face not only our complicity in ecological exploitation, but also our creative responsibility in imagining a more sustainable future.

Bummi's and Carole's stories are deeply entangled with these predicaments, as they are propelled forward by the neo-liberal and capitalist economies of post-colonial Nigeria and contemporary Britain alike. Precariousness, for these women, is not an inner experience – as for Ada – but a lived reality. Bummi grows up under the hazardous living conditions of the Nigerian oil fields. After her father's untimely death in an illegal refinery, she and her mother leave their home in the periphery and head for the capital city, Lagos:

> and they fled the orange gas flares of the refineries burning twenty-four hours a day into the humid skyline for hundreds of miles
> they fled the toxic fumes that made breathing the very air difficult because to inhale deeply was to die slowly
> they fled the acid rain that made the water undrinkable
> they fled the oil spills poisoning the crops, the diseased fisheries in the soupy creeks, the fishing baskets lifted out of the water congealed with gummy black oil.
>
> (Evaristo 2020, 160)

The narrative description of the environment in the novel's stream of consciousness mode foregrounds the effects of the petrol industry on the child's lived experience. The scope of this pollution thus becomes immediately relatable, even palpable for the reader; when Bummi and her mother "flee" from these dire living conditions, this initially seems like a relief. However, Lagos is not a hospitable place either. In the capital city, Bummi's mother finds work at a sawmill, where she, too, is killed in an accident when her daughter is 15. By various lucky twists of fate, Bummi earns a university degree in mathematics and emigrates to London with her loving husband Augustine. In the "Mother City", the couple hope to find a better future for themselves and for their daughter Carole. When Augustine dies from a heart attack, Bummi acquires money to found her own cleaning business.

This is more than sheer coincidence: at the height of her career, Bummi comes full circle with the origins of her own life. Her early life was propelled forward by environmental pollution; now, in a dream, past, present, and future merge into one beautiful moment of saving Earth Mother Gaia, which is worth quoting at length:

> she dreamed of employing an army of women cleaners who would set forth across the planet on a mission to clean up all the damage done to the environment
> […]
> she imagined them all descending in their millions on the Niger Delta and driving out the oil companies with their mop and broom handles transformed into spears and poison-tipped swords and machine-guns
> she imagined them demolishing all the equipment used for oil production, including the flare stacks that rose into the skies to burn the natural gas, her cleaners setting charges underneath each one, detonating from a safe distance and watching them being blown up
> she imagined the local people cheering and celebrating with dancing, drumming and roasted fish
> […]
> Afterwards, she imagined legions of singing women sifting the rivers and creeks to remove the thick slicks of grease that had polluted them, and digging up the land until they'd removed the toxic sublayers of soil
> she imagined the skies opening when the job was done and the pouring of pure water from the now hygienic clouds for as long as it took for the region to be thoroughly cleansed and replenished.
>
> (2020, 170–171)

In this lucid dream, Bummi precariously balances upon a critical-creative edge. Lamenting the environmental pollution of her Nigerian home, she also creatively imagines an "army of women cleaners" rising to battle the destructive forces of the petrol industry, liberating the suffering people, cleaning the natural environment, and celebrating the "now hygienic" new earth. Bummi's dream offers an anti-capitalist critique of global industries and, by extension, of the postcolonial and neo-liberal economy.

It is interesting to note that Bummi finds a way to exploit this capitalist economy for her own green purposes. In continuation of this story, her daughter Carole, too, earns a degree in mathematics and becomes a successful banker in London. Moreover, the lives of mother and daughter significantly overlap: "by the time Carole began her banking career in the City, Bummi had a staff of ten" (177). Whether Carole embraces her mother's dream for a more sustainable future, however, remains unclear. If anything, Carole seems to move smoothly along in Great Britain's capitalist economy, as may be exemplified in a passage describing her daily commute to work:

> Carole
> walks through Liverpool Street station with its inter-galactic glass and steel ceiling propped up by towering Corinthian columns
> she's headed for the escalators and the soaring windows that let in a holy glow of morning light.
>
> (Evaristo 2020, 113)

This passage not only privileges Carole's inner experience, it also exemplifies the ambition and upward movement implied by a capitalist economy: the narrative gaze is directed up to the "inter-galactic glass and steel ceiling", raised along the "Corinthian columns", while Carole's own path is "headed for the escalators and the soaring windows". The promise of her success and a better future reverberates in the "holy" glow of the morning light.

Like her mother, Carole is extremely resilient in overcoming the traumatic experiences of her youth and equally resourceful in managing and imagining the future success her parents had hoped for when they migrated to London. Mother and daughter have moved from a presumed "periphery" into the "centre", the "City" with a capital C: not only the British capital, but also the financial capital.

A similar movement from periphery to centre occurs in Otoo's novel, *Ada's Room*. Totope, the place of Ada's first incarnation and hence the beginning of her story, is located in the periphery of the Ghanaian capital city, Accra, and at the coast. While it was once a site of colonial invasions, it is now threatened by rising water levels. And yet, *Ada's Room* remains strikingly silent on environmental concerns. Only Ada's half-sister Elle is introduced as an ecologically aware person (Otoo [2020] 2023, 200).

Instead, Ada's presence integrates a sense of precariousness via experiences of sexism and racism, of postcolonialism and capitalism, which – returning to *Girl, Woman, Other* – finds a parallel in Bummi's and Carole's stories. Regardless of her education and success, Bummi complains that "people viewed her through what she did (a cleaner) and not what she was (an educated woman)" (Evaristo 2020, 167). Carole, an Oxford graduate and successful banker, braces herself before every meeting:

> she's thinking he'd better not look at her as if she should be attached to a trolley bearing flasks of coffee, assortments of teas […] she's used to clients and new colleagues looking past her to the person they are clearly expecting to meet.
>
> (117)

Even as their stories claim centre-stage, their experiences as women-of-colour in London as much as in Berlin reveal lingering prejudices and discrimination, thus conjuring a sense of utmost precariousness and blurring of the stereotypical contrast between "periphery" and "centre". In this movement, the literary trope of motherhood, too, gains an additional precariousness: while Bummi strives to uphold Nigerian family traditions, she "did not foresee the long-term negative impact of her daughter going to the famous university for rich people" (150), culminating in the realization – whether true or false – that "her daughter would soon belong completely to them" (159), the British. Regardless of the diachronic and synchronic continuities in the experiences of mother and daughter, then, there remains an inevitable and yet, for the first time, an enabling element of precariousness: the possibility to choose a new way forward.

As shown, the critical-creative edge of motherhood has offered new perspectives on the temporal-historical as well as on the spatial-ecological repercussions of assessing stories of the past and of imagining stories of the future. As each of these stories translates into the other, they jointly conjure a new precarious edge: the writerly and readerly tradosphere of "the place *that* we live" (Bennett 2004, 365; emphasis added), to be further explored in the following section.

The critical-creative edge as tradosphere

The critical-creative edges of women's writing and of (literary) motherhood encourage writers and readers alike to engage in acts of translation from past into future. The ensuing momentary present delineates an intermediary tradosphere, which fosters a multiplicity of options to think backward and forward at the same time. These acts of translation occur in a direct relation between embodied human experience and the material more-than-human environment. In these ways, the two novels, *Girl, Woman, Other* and *Ada's Room*, perform an inherent act of translation from experience into (writerly) imagination, into language, into words on the page. However, human language is no more than a "prosthetic tool" for expressing the writer's, the characters', or the readers' "psychophysiological involvement" with their world (Scott n.d., n.p.).

Instead literature calls for creative reading (see Attridge 2017, 117), in which readers engage in an act of re-translating the written text into their own imagination and relating it to their own experience. A reader is thus drawn into the text's ecology (Zapf 2016), and reading becomes an eco-translational practice of habitation. In this process, literary texts come to partake, as stories and material props, in a literary ecology, i.e., in "the story (logos) of the place *that* we live (oikos)" (Bennett 2004, 365; emphasis added).

These conundrums may be exemplified when, in the middle section of *Ada's Room*, Ada's soul dissolves into white emptiness. This is illustrated by four blank pages with only one brief line, placed in the middle: "an awful lot of white | and then actually nothing more | and that for all eternity" (Otoo [2020] 2023, 123–125). As the white pages materialize eternity, they at once play on typical associations with the colour "white", while – in conjunction with the black ink representing Ada's singular experience of dissolving before she will be reborn – the novel may imply a subtle commentary on cultural hegemonies.[4] In Evaristo's *Girl, Woman, Other*, readerly immersion may appear in the writer's adaptation of the modernist stream of consciousness technique with its pervasive lack of punctuation and refusal of spelling conventions. In the words of one (white, male) reviewer:

> Novella-length chapters draw us deep into the lives of 12 women of various backgrounds and experiences. There's nothing forced about the virtual exclusion of white characters from this novel; they have simply been shifted to the periphery, relegated to the blurry sidelines where black characters reside in so much literary fiction written by white authors.
>
> (Charles 2019, n.p.)

From my personal, and necessarily limited vantage of a white reader, I share this experience of being drawn into the characters' stories; I also feel compelled to re-evaluate my own experience of being a woman, a writer, and a mother, to the effect that I begin to rethink my own immersion in the literary tradosphere in my roles as critical reader, as literary scholar, and as a fellow writer who shares the experience of precariousness and who bonds with literary mother and sister figures in seeking creative expressions for planetary habitability.

In this sense, it seems crucial to acknowledge that this tradosphere challenges us to ask ourselves, in line with Anna Lowenhaupt Tsing (2021): "what if […] precarity is the condition of our time or, to put it another way, what if our time is ripe for sensing precarity?" (20). Indeed, the contemporary experiences of precariousness are manifold. On the pivotal edge of critique and creativity, their literary renderings in the contemporary novels *Ada's Room* and *Girl, Woman, Other* have challenged me – as well as, hopefully, other writers and readers of any gender or colour – into a critical awareness of, but also a creative thinking beyond, lingering prejudices and uneven privileges. In this very spirit, Evaristo's *Girl, Woman, Other* is dedicated

> [to] the sisters & the sistas & the sistahs & the sistren
> & the women & the womxn & the wimmin & the womyn
> & our brethren & our bredrin & our brothers & our bruvs

> & our men & our mandem & the LGBTQI+ members
> of the human family
>
> (2020, n.p.)

Finally, in both *Girl, Woman, Other* and *Ada's Room*, the trope of Gaia has offered critical examples for decolonizing literature in gendered, racial, and ecological terms. Above all, these writings draw us, as readers, into a shared "story of the place that we live". Even as this storied ecology pivots on an equally shared experience of precariousness, the ensuing tradosphere may offer a habitable place from which to think back through our (literary) mother figures and to think forward into more sustainable futures for us and for our Earth.

Notes

1 Elaine Showalter (1992) criticizes Virginia Woolf for not being overtly political, and for emphasizing writerly style over social inequalities. At the same time, the private is political in Woolf, and the diverse forms of self-restraint the writer and her women characters impose upon themselves are telling, albeit subtly. Showalter's essay further outlines a development from modernism to the 1970s. As women continue to write in Woolf's shadow, as a dominant mother figure, they "were encouraged to be androgynous, to transcend consciousness of their sex, certainly not to write about it" (215); finally, the 1970s brought about a new tradition in which women articulate "their experiences as bodies" and, in so doing, "these new women writers are anything but sentimental" (218).

2 In her Talk on Literature, given in German, Sharon Otoo delineates her literary influences from English and German literary traditions and diversifies her sense of writerly connectedness in terms of race and gender:

> Somit schreibe ich in der Tradition von Geoffrey Chaucer und Charles Dickens, von Bertold Brecht und Heinrich Böll. […] Ich schreibe im warmen Schatten des nigerianischen Autors Chinua Achebe. […] Ich schreibe in den riesigen Fußstapfen der Schwarzen US-amerikanischen Autorin und Literaturnobelpreisträgerin Toni Morrison […] Ich schreibe in Dankbarkeit für die bahnbrechende Arbeit der Schwarzen deutschen Aktivistin, Pädagogin und Dichterin May Ayim.
>
> (Otoo [2020] 2023, 22–23)

> This is how I write in the tradition of Geoffrey Chaucer and Charles Dickens, of Bertold Brecht and Heinrich Böll. […] I write in the warm shadow of Nigerian author Chinua Achebe. […] I write by following in the huge footsteps of Black US-American author and Nobel Prize winner Toni Morrison. […] I write in gratitude for the groundbreaking work of Black German activist, teacher, and poet May Ayim."
>
> (my translation)

3 While the German term "Schleife" might translate as "loops" of reincarnation, the translator's choice of "orbit" congenially spotlights Ada's circling around a centre that remains perpetually out of reach. This essay spins off Ada's incarnations; it will, therefore, stick to the term "loops" rather than "orbits".

4 Interestingly, similar colour codings are employed in Bernardine Evaristo's earlier novel *Soul Tourists* (2006), in which the protagonist Stan, a man-of-colour in contemporary London, decorates his flat in various shades of white, against which his own presence stands out conspicuously.

References

Attridge, Derek. 2017. *The Singularity of Literature*. London: Routledge.

Bennett, Jane. 2004. "The Force of Things: Steps towards an Ecology of Matter." *Political Theory* 32 (3): 347–372.

Bennett, Jane. 2010. *Vibrant Matter: A Political Ecology of Things*. Durham, NC and London: Duke University Press.

Charles, Ron. 2019. "Bernardine Evaristo's 'Girl, Woman, Other' received half a Booker Prize, but it deserves all the glory." *The Washington Post*, October 28. https://www.washingtonpost.com/entertainment/books/bernardine-evaristos-girl-woman-other-received-half-a-booker-prize-but-it-deserves-all-the-glory/2019/10/28/b22212fa-f97a-11e9-8906-ab6b60de9124_story.html

Eliot, T. S. [1919] 2009. "Tradition and the Individual Talent." *Poetry Foundation*, October 13. https://www.poetryfoundation.org/articles/69400/tradition-and-the-individual-talent

Evaristo, Bernardine. 2006. *Soul Tourists*. London: Penguin Books.

Evaristo, Bernardine. 2020. *Girl, Woman, Other*. London: Penguin Books.

Evaristo, Bernardine. 2021. *Manifesto: On Never Giving Up*. London: Hamish Hamilton.

Gaard, Greta, ed. 1993. *Ecofeminism: Women, Animals, Nature*. Philadelphia, PA: Temple University Press.

Gyasi, Yaa. [2016] 2017. *Homegoing*. London: Penguin.

Hustvedt, Siri. [2020] 2021. "A Walk with my Mother." In *Mothers, Fathers, and Others: Essays*, 21–48. New York: Simon and Schuster.

Iovino, Serenella and Serpil Oppermann, eds. 2014. *Material Ecocriticism*. Bloomington and Indianapolis: Indiana University Press.

Levy, Nadine. 2013. "Towards a New Environmental Ethic in Contemporary Feminist Theory." *Hecate* 38 (1/2): 9–23.

Otoo, Sharon Dodua. [2020] 2023. *Ada's Room*. Translated by Jon Cho-Polizzi. New York: Riverhead Books.

Otoo, Sharon Dodua. 2020. *Dürfen Schwarze Blumen malen? Klagenfurter Rede zur Literatur.* [May Black People Draw Flowers? Klagenfurt Talk on Literature] Klagenfurt: Heyn. Accessed May 31, 2024. https://files.orf.at/vietnam2/files/bachmannpreis/202022/otoo_klagenfurter_rede2020_ansicht4_752860.pdf

Otoo, Sharon Dodua. 2022. *Herr Göttrup setzt sich hin*. [Mr Göttrup sits down.] Frankfurt am Main: Fischer.

Plumwood, Val. 1993. *Feminism and the Mastery of Nature*. London and New York: Routledge.

Schulkind, Jeanne. 1985. "Introduction." In *Moments of Being*, edited by Virginia Woolf, 11–24. Boston and New York: Mariner Books.

Scott, Clive. n.d. "Working With the Notion of Eco-Translation." Talk held at the University of Exeter. *Soundcloud*. Accessed March 2, 2023. https://soundcloud.com/university-of-exeter/prof-clive-scott-working-with-the-notion-of-eco-translation

Shange, Ntozake. [1975] 2021. *For colored girls who have considered suicide / when the rainbow is enuf*. New York: Scribner.
Showalter, Elaine. [1972] 1992. "Killing the Angel in the House: The Autonomy of Women Writers." *The Antioch Review: 50th Anniversary Issue* 50 (1/2): 207–220.
Stanford Friedman, Susan. 1987. "Creativity and the Childbirth Metaphor: Gender Difference in Literary Discourse." *Feminist Studies* 13 (1): 49–82.
Tsing, Anna Lowenhaupt. 2021. *Mushroom at the End of the World: On the Possibility of Life in Capitalist Ruins*. Princeton, NJ: Princeton University Press.
Woolf, Virginia. [1928] 1945. *A Room of One's Own*. London: Penguin Books.
Woolf, Virginia. [1942] 1985. "A Sketch of the Past." In *Moments of Being*, edited by Jeanne Schulkind, 61–160. Boston, MA and New York: Mariner Books.
Woolf, Virginia. [1925] 1992a. *Mrs Dalloway*. Oxford: Oxford University Press.
Woolf, Virginia. [1931] 1992b. *The Waves*. Oxford: Oxford University Press.
Woolf, Virginia. [1928] 1993. *Orlando*. London: Penguin Books.
Woolf, Virginia. (1942) 2012. "Professions for Women." In *The Death of the Moth and Other Essays*, edited by Leonard Woolf. *Project Gutenberg Australia*, October. https://gutenberg.net.au/ebooks12/1203811h.html#ch-28
Woolf, Virginia. [1927] 2019. *To the Lighthouse*. London: Vintage.
Zapf, Hubert. 2016. *Literature as Cultural Ecology: Sustainable Texts*. London: Bloomsbury.

10 Earth is *Oikos*

Peter Sanger on the vulnerability of the biosphere as life's home

Leonor María Martínez Serrano

(Non)human life's home

Canadian poet and critic Peter Sanger (b. 1943) is the author of a prolific body of work, including poetry collections, essays, and edited books. This chapter looks into his ecological thinking as put forward in an essay titled *Oikos*, originally delivered as an address at the Nova Scotia Agriculture College on May 4, 2012, on the occasion of the award of an Honorary Doctorate of Laws from Dalhousie University, and published as an elegant chapbook by Gaspereau Press in 2014. *Oikos* is the ancient Greek word for "home", which in Sanger's thinking is synonymous with "Earth, planet or biosphere" – the home life has patiently built for itself – and it appears to be indebted to Heidegger's thinking on nature, poetry, and dwelling. At a time of alarming environmental degradation that is seriously compromising the habitability of the Earth, and the persistence in time not just of *Homo sapiens*, but also of a myriad of nonhuman species, it is wise to listen closely to what poet-thinkers tell us of value about the moral imperative to protect the Earth and its life forms.

Sanger is not alone in thinking of the biosphere as a fragile place that needs our protection and respect, as we are not its owners – "*we don't own what we know*" (quoted in Bringhurst 2006, 270) – as ecopoet Don McKay elegantly puts it in just six syllables. He is part of a long tradition that comprises the pre-Socratic poet-philosophers, the ancient Chinese poets and Buddhist sages, the Transcendentalists R. W. Emerson and H. D. Thoreau, Martin Heidegger, Aldo Leopold (1949; 1991a), Freya Mathews (2003), David Abram (1996; 2010), Jane Bennett (2010), and a host of contemporary ecopoets and thinkers intent on thinking beyond "bifurcations of nature and culture" (Cohen and Duckart 19) and on denouncing the ferocious, shameless exploitation of the planet, our shared home, with different degrees of belligerence and explicitness. In the spirit of Serenello Iovino's (2010) concept of "non-anthropocentric humanism", they claim that there is another way to live on Earth, with a sense of duty

DOI: 10.4324/9781003383598-14

and responsibility, that life (not *Homo sapiens)* is the raison d'être of the biosphere, and that it is biocentrism rather than anthropocentrism that we should embrace. In light of these insights, this chapter argues that the intellectual roots of Sanger's eco-thinking can be traced back to a centuries-old tradition sensitive to the vibrancy of matter, the livingness of the Earth, and the indivisibility of body and environment, which Stacy Alaimo (2008) has aptly called "transcorporeality".

One of those eminent ancestors was the Greek philosopher and physicist Empedocles, who "argued that all matter consists of four elements in shifting combination: earth, air, fire, water" (Cohen and Duckart 2015, 2). Attracted by love (*philia*) and pulled apart by strife (*neikos*), the elements were conceived as being the "primal 'roots'" (2), or the common substratum to all things making up the cosmos. *Cosmos* is, incidentally, the Greek word for "order" and "harmony". The word for "element" in Aristotle's Greek was *stoikheion*, which was then rendered as *elementum* in Latin and as *Urstoff* or "primordial substance" in German (Cohen and Duckart 2015, 7–8). Larger than atoms in a wide range of manifold manifestations, "[t]he elements are matter's kinetic syllabary" (8) and "the perceivable foundations of which worlds are composed, the animated materialities with and through which life thrives" (13). With great intellectual alertness, Empedocles named them "*rhizomata*, tangling roots" (8), for they were the *arché* or irreducible origin of what-is. In the Empedoclean account of the world, written in the alphabet of the elements, matter is "inherently creative, motile, experimental, impure" (3). He "believed human bodies to be microcosms: composed of all four 'roots'", but he was not oblivious to the fact they are not "the measure of the world" (12), just one of the many forms that matter may assume in endless, protean combinations. By embracing "ecomaterialism" (5), what Cohen and Duckart (2015) seek is "a disanthropocentric reenvisioning of the complicated biomes and cosmopolities within which we dwell" (5), since humans cannot be dissociated from the world of which they are a part. They argue that humankind appears to have forgotten that matter is "a precarious system and dynamic entity", which has resulted in a process of "relentless objectification" that has ultimately transformed not only nature, but also "even humans into expendable resources" (5). Cohen and Duckart's "environmental agentism" posits nature as being an "active force, unlooked-for partner", whilst the elements are "open portals to spaces that pulse with inhuman life" (2015, 6). The web of life is a complex, dynamic, endlessly changing vortex pulsating with earth, air, water, and fire. In pondering the relevance of the word "ecology" in *Oikos*, Sanger, like many of his predecessors, reminds 21st-century readers that *Homo sapiens* is not the measure of all things, but a thread in the texture of being instead.

Semantic archaeology

In a journal entry dated April 2, 1852, Henry David Thoreau (1906) wrote: "Man is altogether too much insisted on. The poet says the proper study of mankind is man. I say study to forget all that – take wider views of the universe. That is the egotism of the race" (369). The Transcendentalist thinker was well aware of the egotism of our species and, most importantly, he was sensitive to the mutual entanglements of all life forms in the mesh of things we call "world", "cosmos", or "universe". Looking at humankind as part of a vast network of relationships was, therefore, much more interesting to him than just observing men and women in their daily commerce. There is abundant evidence in Sanger's work of a lifetime that he turned to the more-than-human world to investigate the great concerns of humanity – to shed light on reality and the place occupied by *Homo sapiens* within the larger scheme of things. A case in point is *Oikos*, a brief yet enlightening essay on ecology and the vulnerability of the Earth which remains completely eloquent in its brevity. Sanger thinks of "ecology" as being a word "whose definition in the widest possible, accurate, historical sense should be of crucial interest to us as living human beings" (2014, 5). Hence the very first thing he does in exploring his object of inquiry is to look up the word "ecology" in an authoritative dictionary and think about the implications of putting a word "into the general circulation of dictionary definition". He writes thus:

> [The word "ecology"] does not appear in the great, multi-folios of the *Oxford English Dictionary* (which took over forty years to complete) of 1928. The word does appear in the *Concise Oxford Dictionary*, fourth edition, of 1951 where it is defined as "a branch of biology dealing with living organisms' habits, modes of life and their relations to their surroundings".
>
> (2014, 5)

It does come as a surprise that such a massive, comprehensive dictionary as the *Oxford English Dictionary* (*OED*), a work in progress for over four decades, should have omitted the word "ecology", particularly on account of the reality it designates, which should be of everyone's concern for obvious reasons. As members of the species *Homo sapiens* are also living organisms, have distinct modes of life, and relate to their environment in specific manners, ecology as a branch of knowledge concerned with the study of living forms should be of interest to all of us. At any rate, every time a word is "put into the general circulation of dictionary definition", explains Sanger, "then we can know that there has been a major shift in human knowledge, concern, in human self-definition and ultimately responsibility" (2014, 5). As British ecocritic Jonathan Bate (2015) observes in "The Economy of Nature", the word "ecology" (from the German *oekologie*) was "coined in 1866 by the German zoologist Ernst Haeckel and defined more fully by the same scientist in 1870" (77). The definition provided by Haeckel reads as follows: "[T]he body of knowledge concerning the economy of nature – the investigation of the total relations of the animal both to its inorganic and to its organic environment" (77). Though separated by a time span of 80 years, Haeckel's definition of 1870 and that of the *OED*'s of 1951 both emphasize the notion of relationship between any specific life form and its environment. In a subtle way, both definitions gesture towards the fact that the biosphere is a vast constellation of connections, a gigantic network of relationships where everything is connected to everything else.

Nature, biosphere, and *oikos* as the root word behind "ecology" are closely related to each other. In his essay, Sanger's analytical mind seeks to clarify the meaning of the word "ecology" by looking at its etymology, that is, he conducts a thorough exercise in semantic archaeology. As there is a wealth of knowledge buried within words that tends to be taken for granted, looking closely at their roots is like opening a door to step into a realm of primordial meanings. Apropos the etymology of "ecology", Sanger notes the following:

> The word "ecology" derives from the Greek word *oikos*, which means "house." The second part of "ecology" (the "logy" part) derives from the word logos, meaning "word" or "reason", and those meanings, in turn, lie at the heart of both early Greek and Christian philosophy. In other words, "ecology" means literally, according to its semantic roots, "house reason" or "house words", and construal of "ecology's" full meaning must contain, at the very least, recognition that the biological is our home. […] "Ecology" is "earth word", "earth reason."
>
> (2014, 6)

"The biological is our home" (6), writes Sanger succinctly. In other words, *oikos* is Earth in its entirety, and Earth is home to the human and nonhuman alike. Earth as *oikos* is a vulnerable place that it is everyone's responsibility to protect at all costs. In fact, Sanger is aware of how fragile and vulnerable everything we love can be. According to the *Oxford English Dictionary* (*OED*), "vulnerable. adj." means "that may be wounded; susceptible of receiving wounds or physical injury" (*OED* 2024, http://www.oed.com). *Homo sapiens*'s place resides next to other life forms on Earth, not over or above the rest of creation. However, in the creation myth recounted in the biblical Book of Genesis, man is commanded to take dominion over the world. The story is familiar enough. As rendered from the original Hebrew into English, this is how it reads:

> God created man in his own image,
> in the image of God created he him;
> male and female created he them.
> And God blessed them,
> and God said unto them,
> Be fruitful, and multiply, and replenish the earth,
> and subdue it:
> and have dominion over the fish of the sea,
> and over the fowl of the air,
> and over every living thing that moveth upon the earth.
> (Bringhurst and Zwicky 2018, 10)

"Ecology", from *oikos* and *logos*, is literally the "study of the house", and yet the definition of *oikos* remains a challenge. In her classic study *What Is Nature?*, Kate Soper (1995) explores the multiple meanings of the word "nature". In contradistinction to "culture", "nature" primarily designates "everything which is not human and distinguished from the work of humanity" or, put simply, absolute "'otherness' to humanity" (15). This is the basic sense of the word "nature", but, in Soper's view, nature is also "that totality of being of which we in some sense conceive ourselves as forming a part" (21). Though it might seem contradictory, "nature" simultaneously refers to "both that which we are not *and* that which we are within" (21; original emphasis), a statement that is expressive of the dualistic conception of existence central to our western mindset. The Cartesian mind versus matter dichotomy is deeply ingrained in the western conceptualization of the universe. It parses reality into animate and inanimate entities, into *res cogitans* (mind, spirit) and *res extensa* (body, inert matter). This partition of the sensible has been challenged by philosophers like Freya Mathews (2003), who, in positing panpsychism, endorses Leibniz's and Spinoza's intuition "that everything is woven from the same skein, that everything informs everything else, and hence that 'mind' in some sense must be intermingled with 'matter'" (3). Drawing on their "intimation of interpermeation – of everything participating in everything else, of mutual holding and enfolding at the structural base of existence" (3), Mathews embraces a view of the world that "implicates mind in matter and matter in mind") and a form of "nonduality or mind-matter unity" (27) that transcends the Cartesian dualistic conception of mind and matter as "categorically distinct substances" (26).

David Abram (2014) makes use of the term "The Commonwealth of Breath" (313), to refer to all Earth dwellers who are implicated in a shared semiotic-material substratum, holding together the human and nonhuman realms. In fact, the notion of a "more-than-human world" is put into the circulation of ecocritical discourse for the first time in Abram's (1996) study *The Spell of the Sensuous*, where he writes: "we feel ourselves enveloped, immersed, caught up within the sensuous world. This breathing landscape is no longer just a passive backdrop against which human history unfolds, but a potentized field of intelligence in which our actions participate" (260). Acknowledging breath as the common substratum to all earthbound beings and entities, Abram takes us back to a fundamental insight – i.e., that both humans and nonhumans are all part of a breathing Earth that centuries of scientific research have reduced to sheer exteriority. As Abram (2010) puts it, "After three and a half centuries spent charting and measuring material nature as though it were a pure exterior, we've at last begun to notice that the world we inhabit [...] is alive" (158). Humanity has come to the realization that the world where we dwell is alive with "intermingling agencies and forces that persist and change over eons", mutually entangled in "a constant process of shared becoming" (Iovino and Oppermann 2014, 1).

Mens agitat molem (mind moves matter)

The intuition that the Earth is *oikos* and a living entity endowed with some form of mind can be traced back to earlier philosophers and poets – to Abram's and Sanger's intellectual ancestors. In this connection, in what might be considered the palpitating heart of his essay, Sanger chooses to focus on Virgil's panpsychic ecologism *avant la lettre* as manifested in "one of the central texts in Western culture, [...] *The Aeneid*" (Sanger 2014, 9). The Roman poet's conceptualization of Earth as a place endowed with sentience and consciousness is discernible in a passage from Book VI where Anchises, Aeneas's father, "full of the knowledge that his life in the afterlife has brought him, tells his son [...] something of the structure of the cosmos" (9). The moment seems to be propitious for the father to share such a momentous revelation with his son, surrounded as they are by throngs of shades on the banks of the Lethe awaiting their reincarnation.[1] Virgil's Latin text (ll. 6.724–6.732) reads thus in the original hexameters composed by the poet in imperial Rome:[2]

> Principio caelum ac terras camposque liquentis lucentemque globum lunae Titaniaque astra spiritus intus alit, totamque infusa per artus mens agitat molem et magno se corpore miscet. inde hominum pecudumque genus, vitaeque volantum, et quae marmoreo fert monstra

> sub aequore pontus. igneus est ollis vigor et caelestis origo seminibus, quantum non noxia corpora tardant terrenique hebetant artus moribundaque membra.
>
> (Virgil 1916, 556)

In Dryden's 1697 English translation, which Sanger scrutinizes in *Oikos*, the passage reads:

> Know first, that heaven and earth's compacted frame,
> And flowing waters, and the starry flame,
> And both the radiant lights, one common soul
> Inspires and feeds and animates the whole.
> This active mind, infused through all the space,
> Unites and mingles with the mighty mass.
> Hence men and beasts the breath of life obtain,
> And birds of air and monsters of the main.
> The ethereal vigour is in all the same;
> And every soul is filled with equal flame.
>
> (Virgil 1909, 235–236)

A close analysis of this excerpt from the *Aeneid* shows it is not devoid of aesthetic and intellectual rewards. As Sanger notes, the original Latin "*mens agitat molem*" (line 6.727) is rendered into English as "active mind, infused through all the space, / Unites and mingles with the mighty mass", asserting also the common breath shared by "men... beasts... birds of air... monsters of the main" (2014, 11). In translation, words are deliberately selected, not a random choice. In translating "*mens agitat molem*" like this, Dryden wishes to emphasize (1) that there is a universal principle uniting all life forms into a unitary whole, (2) that matter, far from being inert or dead, is vibrant stuff, as argued by Barad (2007) and Bennett (2010), and (3), that the world has a mind of its own, discernible in a vast spectrum of life forms. Prominence is also given to the common breath shared by men, beasts, birds, and monsters, a notion that Abram would elegantly capture through the term "Commonwealth of Breath" (2014, 313) some centuries later. It is breath that suffuses the world and its dwellers, and in so doing reunites the Many into the One of which they are an expression. This is Sanger's critical analysis of Dryden's deliberate word choice:

> In bleaker, modern English, "*Mens agitat molem*" means "Mind stirs the soil." Dryden's expansionary paraphrase turns the laconic Latin into poetry, as Dryden understood it. But his paraphrase is also an ontological redress intended to reattach the English word 'mind' to Platonic and

> Pythagorean traditions. Therefore, Dryden's version emphasizes that a metaphysical '*mens*' animates a physical "*molem*".
>
> (Sanger 2014, 11; original emphasis)

It is at this point that Virgil's, Dryden's, and Sanger's thinking on the universal breath suffusing the Earth in its entirety ties up with Abram's postulates in *Becoming Animal* (2010), where the ecophilosopher seeks to raise awareness that human beings are "enworlded", embodied minds, and, as such, are "part of" – not "apart from" – the more-than-human world, where we are participants in a completely sensuous immersion together with other sentient, ensouled beings. In so doing, Abram highlights "our interbeing with the earth" and seeks to instill in us "a new humility in relation to other earthborn beings" (2010, 3). What he accomplishes is nothing less than the construction of an earthly cosmology and sensuous physics of being in the face of the piecemeal destruction of the Earth in the Anthropocene and our species' increasing detachment from the living world. We have forgotten "our carnal inherence in a more-than-human matrix of sensations and sensibilities" (1996, 22), writes Abram. He acknowledges the indivisibility of reality, the existence of sentient beings and intelligences other than our own, and our animal bodies' yearning for the larger body of the biosphere. Ultimately, the Earth is revealed to be a huge breathing body populated by "ensouled" bodies where mind and matter intermingle. The common material substratum to all human and nonhuman beings connects them across boundaries and brings them into complex material-semiotic relationships. Aware of the porosity of the membrane separating us from the world and the "sensorial reciprocity between our breathing bodies and the bodily terrain" (1996, 267), Abram sees perception itself as a dialogue between our sentient bodies and the larger body of the Earth we are inextricably linked to. There is no hierarchy then, with our species on top, but rather a democracy or horizontality of being: "With the other animals [...] we're all implicated within this intimate and curiously infinite world [...] within the same big interior. Inside the world" (Abram 2010, 158).

Abram must have read Aldo Leopold's writings, as there are striking similarities between their respective reflections on the Earth and ecology. In an early essay titled "Some Fundamentals of Conservation in the Southwest" (1923), Leopold conceives of the biosphere as a breathing body endowed with a consciousness of its own. He argues that "[m]any of the world's most penetrating minds have regarded our so-called 'inanimate nature' as a living thing" (1991a, 94). In this view of an "organized animate nature" (94), he claims, it is possible "to regard the earth's parts – soil, mountains, rivers, atmosphere, etc. – as organs, or parts of organs, of a coordinated whole" (95) and to ascribe to earth "that invisible attribute – a soul, or consciousness – which [...] many philosophers of all ages, ascribe to all living things and aggregations thereof" (95). This is exactly the same *mens* or mind that Virgil has Anchises explain to Aeneas in Book VI of the *Aeneid* during his sojourn in Hades. The same indivisibility that Virgil senses in what-is is discernible in Leopold's conclusion to his essay: "in our intuitive perceptions, [...] we realize the indivisibility of the earth [...] and respect it collectively not only as a useful servant but as a living being" (1991a, 95). A most subtle thread thus links the thinking of Virgil, Dryden, Leopold, Abram, and Sanger.

If *Homo sapiens* is part of something larger than us, then it is our responsibility not to commodify and shamelessly exploit the nonhuman world, but to relearn to attend and relate to it in a different way instead. As Heidegger (1993) observes in memorable words in his "Letter on Humanism": "Man is not the lord of beings. Man is the shepherd of Being" (245). It is with ethics, and not ontology, that we should begin our attempts to understand the world and the place we occupy within it. Upon closer inspection, Heidegger seems to imply, epistemology, ontology, and ethics are interdependent. In "The Origin of the Work of Art", he claims that "[l]anguage itself is poetry in the essential sense" (1971, 72), meaning that the primordial goal of language is "projective saying", i.e., the naming of the beings and things dwelling on Earth, thus bringing being into the open for the first time. As he puts it, "[l]anguage, by naming beings for the first time, first brings being to word and to appearance. Only this naming nominates beings *to* their being *from out of* their being" (1971, 71; original emphasis). In this train of thinking, poetry is conceived as being a force that enunciates being into existence; it is "the saying of the unconcealedness of what is" (71). The potency of poetry to unveil being makes it a crucially relevant discourse nowadays amidst the Anthropocene. It might teach us to pay attention to the nonhuman. More specifically, ecopoetry and ecophilosophy may advance positions of realistic hope and contribute to the flowering of a new, relational ontology where the nonhuman is perceived as what it is – part of the very same skein the human is made of.

Towards the cosmocene: "We are kindred"

In a short article titled "The Anthropocene", Paul J. Crutzen and Eugene F. Stoermer (2000) highlight "the central role of mankind in geology and ecology by proposing to use the term 'Anthropocene' for the current geological epoch" on account of the growing "impacts of human activities on earth and atmosphere" (17). The onset of this new geological epoch can be traced back to the latter part of the 18th century, coinciding with the first stages of industrialization on a massive scale. They argue that ever since the beginning of the Industrial Revolution, "the global effects of human activities have become clearly noticeable" (17) and so they predict that "mankind will remain a major geological force for many millennia, maybe millions of years, to come" (18). As environmental historian Jason W. Moore (2016b) has convincingly argued, the root cause of the current planetary emergency humankind is facing lies in the dynamics of extractive and predatory capitalism, "a system of getting nature […] to work for free or very low-cost" (11). According to Moore, the Capitalocene is "a multispecies assemblage, a world-ecology of capital, power, and nature" (2016a, xi). As a productive conceptual tool that underscores the entanglements of society and nature, the Capitalocene conceives of human organization as "utterly, completely, and variably porous within the web of life" (Moore 2016b, 5) and envisions "humanity-in-nature" (6). The way western industrialized societies relate to nature cannot be divorced from prevalent capitalism, which has increasingly impacted the Earth over time: "imperialism and global trade; plantations and industrial-scale agriculture; factories and the carbon economy […]; the twentieth century's acceleration of consumerism" (Parham 2021, 2).

Pondering the connections between "ecology" and other cognate words such as "economy", Sanger writes: "An economy divided against an ecology cannot stand" (2014, 11); it is simply self-destructive, self-suicidal. Though he does not explicitly make use of the terms "Anthropocene" or "Capitalocene" in *Oikos* to refer to the current environmental crisis, he argues that "[p]olitics, resource extraction and manipulation, multi-national business, all matters of war and peace can only be conducted in unreasonable, illogical, self-destructive and self-contradictory ways if earth reason, the house reason of ecological economy, is denied or ignored" (7). Owing to the current practices of industrialized, capitalist societies, "[w]e have reached the moment in technological possibility where 'ecocide,' 'genocide' and 'suicide' have become synonymous terms" (7), claims Sanger. There is room for realistic hope, though. Geographer Jamie Lorimer (2015) has conjectured that the Cosmocene "would begin when modern humans became aware of the impossibility of extricating themselves from the earth and started to take responsibility for the world in which they lived" (4).

"A post-Natural epoch of multispecies flourishing" (Lorimer 2015, 4) after the Anthropocene, the Cosmocene posits that wildlife is "not Nature" but rather is everywhere, "in our bodies, our homes, and our cities, as well as in the familiar territories that concern conservationists" (180). It, therefore, calls for the redefinition and repositioning of human beings amidst "a multitude of multispecies entanglements" (181), as well as an ethos of "responsibility to distant human and nonhuman others" (193).

Sanger is not alone either in denouncing the causes behind the current environmental crisis we are faced with. Contemporary ecopoets and ecothinkers have also thought deeply about the consequences and impact of human action in the Anthropocene. Thus, in "The Mind of the Wild", a perceptive essay collected in *Learning to Die* (2018), Robert Bringhurst has described the Anthropocene as "an episode in geologic time during which the wild is subverted by the tame" and as "a geological *event*: a momentary though possibly momentous blip in the earth's biography" (2018, 17; original emphasis). According to Bringhurst, humans' hubris and the logic of overexploitation have taken us to a concerted attempt at destroying the biosphere, put in train by "overbreeding, overbuilding, overexploiting, overhunting, overfishing, and by relentlessly consuming fossil fuel" (19). The commodification of nature and its resources has resulted in a perverse logic – "a calculative logic utterly oblivious to the life of the land" (Abram 1996, 267) – whereby the Earth is seen as something to be exploited to suit humans' needs, even if such exploitation goes beyond all reasonable proportion and seriously compromises the continuity of the biosphere. And yet, the Earth is "not a portfolio of resources for us or our species to buy and sell or manage or squander as we please" (Bringhurst 2018, 12), contends the Canadian poet-philosopher in words that recall Leopold's essay "Wherefore Wildlife Ecology" (1947), where the conservationist argues that the land "is something more than a breadbasket" (1991b, 337). Rather, it is "a community of which [*homo sapiens*] is only a member, albeit now the dominant one" (337). The results of anthropogenic action are irreversible, notes Bringhurst, if we are to believe the conclusions drawn by the scientific community based on massive evidence of utter devastation of ecosystems, terraforming, species extinction, climate change, deforestation, rising sea levels, wildfires, the melting of the poles, acidification of the oceans of the planet, and a myriad of attendant woes resulting in social and environmental injustice.[3] Humans cannot be separated from the land. If we make it uninhabitable, we are damaging the *oikos* that sustains life. In Bringhurst's words: "The earth's life is much larger than our own lives, but our lives are part of it. If we take that life, we take our own" (2018, 12).

Sanger is also aware of the interconnectedness of the human and the nonhuman. In *Oikos*, he comes close to positing a non-anthropocentric humanism like the one postulated by material ecocritic Serenella Iovino in a 2010 paper titled "Ecocriticism and a Non-Anthropocentric Humanism: Reflections on Local Natures and Global Responsibilities", published four years earlier than Sanger's *Oikos*. Iovino's proposal of "an extended, non-anthropocentric, humanism" is one that embeds humanism in an ecological paradigm marked by an "ethical 'culture of co-presence'" (32) and an awareness of "our ecological interdependence" (33). She thus endorses Virgil's "*mens agitat molem*" and Dryden's "active mind, infused through all the space, / [which] Unites and mingles with the mighty mass" (Sanger 2014, 11). In other words, Iovino highlights the common substratum uniting all life forms, whilst decentering *Homo sapiens* from the overall picture and advocating the need to cultivate a form of relating to the Earth with a renewed sense of responsibility. It is no coincidence that Sanger's essay should close on an optimistic note, saying: "once upon a time in old English usage nature" was called "kind. [...] [W]e are kindred in kind" (2014, 14).

In this connection, Sanger's work on the oral literatures of the First Nations of North America is a significant accomplishment. In *The Stone Canoe* (2007), he studies two Mi'kmaq texts preserved in manuscript form that had been lost for a long time in the special collections of Acadia University in Wolfville, Nova Scotia. Despite their distinctiveness and uniqueness, they show that among the Indigenous peoples of Canada there are certain commonalities concerning how they perceive and relate to the Earth. One is that the land is sacred, their caretaker and sustainer, the ultimate source of their identity, language, and knowledge. In their *Weltanschauung*, men and women are not the sole possessors of mind, and the land is perceived as having a mind of its own. Its dwellers – human and nonhuman alike – are living beings endowed with sentience and consciousness. As Abram explains in *The Spell of the Sensuous*, among oral peoples language is used not only to communicate with our fellow human beings, but also "to converse with the more-than-human cosmos, to renew reciprocity with the surrounding powers of earth and sky, to invoke kinship even with those entities which, to the civilized mind, are utterly insentient and inert" (1996, 71). Substantial evidence for this claim can be found in their songs, prayers, stories, and myths. The vitality intrinsic to apparently inanimate entities is exactly what Sanger identifies in the two lost Mi'kmaq texts painstakingly transcribed by ethnographer and missionary Silas Tertius Rand from the lips of Susan Barrs back in 1847 and from Old Man Stevens in 1884, respectively. Both stories are among the earliest extant examples of Indigenous Canadian literature recorded in their original language. Both testify to the First Nations' alertness to the livingness of the nonhuman.

Conclusion: breathing Earth

Despite its brevity, Sanger's *Oikos* remains an important contribution to the current debate and global conversation on the unprecedented environmental crisis. As a verbal artefact, it inscribes itself in a larger constellation of texts and in a tradition of eco-thinking that is alert to the vulnerability of the Earth and its life forms. The limpidity of its prose embodies a wealth of meaning and teaches an important lesson. It denounces the impact of human activity on the nonhuman, reminds us of the *mens* animating the Earth, and instills in us a sense that it is possible to reimagine how we as a species relate to the world. That this should be done within the compass of a short text is not a negligible accomplishment. Brevity is often the mark of genius, and Sanger communicates much with few words.

Notes

1 See American poet Mark Strand's (2000) masterful translation of a passage of Book VI of *The Aeneid* where Anchises unsuccessfully tries to embrace his father's shade in Hades; it precedes the passage Sanger briefly analyses in *Oikos*

2 H. Rushton Fairclough's prose translation reads: "First, the heaven and earth, and the watery plains, the shining orb of the moon and Titan's star, a spirit within sustains, and mind, pervading its members, sways the whole mass and mingles with its mighty frame. Thence the face of man and beast, the life of winged things, and the strange shapes ocean bears beneath his glassy floor. Fiery is the vigour and divine the source of those life-seeds, so far as harmful bodies clog them not, nor earthly limbs and mortal frames dull them" (Virgil 1916, 557).

3 Published on November 10, 2021, the "Official Draft of the COP26 Glasgow Agreement" highlights "the urgent need for action" and "[e]xpresses alarm and concern that human activities have caused around 1.1°C of global warming to date and that impacts are already being felt in every region" (United Nations 2021, 2). It also acknowledges that "climate change has already and will increasingly cause loss and damage and, as temperatures rise, impacts from climate and weather extremes [...] will pose an ever-greater social, economic and environmental threat" (7). Though a number of measures are suggested, the agreement seems insufficient if global warming is to be drastically limited 1.5°C by 2100.

References

Abram, David. 1996. *The Spell of the Sensuous: Perception and Language in a More- Than-Human World*. New York: Pantheon Books.

Abram, David. 2010. *Becoming Animal: An Earthly Cosmology*. New York: Knopf.

Abram, David. 2014. "Afterword: The Commonwealth of Breath." In *Material Ecocriticism*, edited by Serenella Iovino and Serpil Oppermann, 301–314. Bloomington and Indianapolis: Indiana University Press.

Alaimo, Stacy. 2008. "Trans-Corporeal Feminisms and the Ethical Space of Nature." In *Material Feminisms*, edited by Stacy Alaimo and Susan Hekman, 237–264. Bloomington and Indianapolis: Indiana University Press.

Barad, Karen. 2007. *Meeting the Universe Halfway: Quantum Physics and the Entanglement of Matter and Meaning*. Durham, NC and London: Duke University Press.

Bate, Jonathan. 2015. "The Economy of Nature." In *Ecocriticism: The Essential Reader*, edited by Ken Hiltner, 77–96. London: Routledge.

Bennett, Jane. 2010. *Vibrant Matter: A Political Ecology of Things*. Durham, NC: Duke University Press.

Bringhurst, Robert. 2006. "Wild Language." In *The Tree of Meaning: Thirteen Talks*, 257–276. Kentville, NS: Gaspereau Press.

Bringhurst, Robert. 2018. "The Mind of the Wild." In *Learning to Die: Wisdom in the Age of Climate Crisis*, edited by Robert Bringhurst and Jan Zwicky, 7–39. Saskatchewan: University of Regina Press.

Cohen, Jeffrey Jerome, and Lowell Duckart, eds. 2015. "Introduction: Eleven Principles of the Elements." In *Elemental Ecocriticism. Thinking with Earth, Air, Water, and Fire*, edited by Jeffrey Jerome Cohen, and Lowell Duckart, 1–26. Minneapolis, MN and London: University of Minnesota Press.

Crutzen, Paul J. and Eugene F. Stoermer. 2000. "The Anthropocene." *The International Geosphere-Biosphere Programme Newsletter* 41: 17–18.

Heidegger, Martin. 1971. "The Origin of the Work of Art." In *Poetry, Language, Thought*. Translated by Albert Hofstadter, 17–76. New York: Harper and Row.

Heidegger, Martin. 1993. "Letter on Humanism." In *Basic Writings*, edited by David Farrell Krell, 217–265. London and New York: Routledge.

Iovino, Serenella. 2010. "Ecocriticism and a Non-Anthropocentric Humanism: Reflections on Local Natures and Global Responsibilities." In *Ecocritical Perspectives on the New English Literatures*, edited by Laurenz Volkmann, Nancy Grimm, Ines Detmers, and Katrin Thomson, 29–53. Leiden: Brill.

Iovino, Serenella, and Serpil Oppermann. 2014. "Introduction: Stories Come to Matter." In *Material Ecocriticism*, edited by Serenella Iovino and Serpil Oppermann, 1–17. Bloomington and Indianapolis: Indiana University Press, 2014.

Leopold, Aldo. 1949. *A Sand County Almanac and Sketches Here and There*. Oxford: Oxford University Press.

Leopold, Aldo. 1991a. "Some Fundamentals of Conservation in the Southwest." In *The River of the Mother God and Other Essays*, edited by Susan L. Flader and L. Baird Callicott, 86–97. Madison, WI: University of Wisconsin Press.

Leopold, Aldo. 1991b. "Wherefore Wildlife Ecology." In *The River of the Mother God and Other Essays*, edited by Susan L. Flader and L. Baird Callicott, 336–337. Madison, WI: University of Wisconsin Press.

Lorimer, Jamie. 2015. *Wildlife in the Anthropocene: Conservation after Nature*. Minneapolis, MN: University of Minnesota Press.

Mathews, Freya. 2003. *For Love of Matter: A Contemporary Panpsychism*. Albany, NY: SUNY Press.

Moore, Jason W. 2016a. "Acknowledgments." In *Anthropocene or Capitalocene? Nature, History, and the Crisis of Capitalism*, edited by Jason W. Moore, xi–xii. Binghamton, NY: PM Press.

Moore, Jason W. 2016b. "Introduction: Anthropocene or Capitalocene? Nature, History, and the Crisis of Capitalism." In *Anthropocene or Capitalocene? Nature, History, and the Crisis of Capitalism*, edited by Jason W. Moore, 1–11. Binghamton, NY: PM Press.

Oxford English Dictionary [*OED*]. n.d. "vulnerable" (adj). https://www.oed.com/search/dictionary/?scope=Entries&q=vulnerable

Parham, John. 2021. "Introduction. With or Without Us: Literature and the Anthropocene." In *The Cambridge Companion to Literature and the Anthropocene*, edited by John Parham, 1–33. Cambridge: Cambridge University Press.

Sanger, Peter. 2007. *The Stone Canoe: Two Lost Mi'kmaq Texts*. Translated by Elizabeth Paul. Kentville, NS: Gaspereau Press.

Sanger, Peter. 2014. *Oikos. An Essay*. Kentville, NS: Gaspereau Press.
Soper, Kate. 1995. *What Is Nature?* Oxford and Cambridge, MA: Blackwell.
Strand, Mark. 2000. "Some Observations of *Aeneid* Book VI." In *The Weather of Words: Poetic Invention*, 75–91. New York: Knopf.
Thoreau, Henry David. 1906. *The Journal of Henry David Thoreau*. Boston and New York: Houghton, Mifflin and Company.
United Nations. 2021. "Official Draft of the COP26 Glasgow Agreement." https://edition.cnn.com/2021/11/10/world/cop26-draft-agreement-full-text-climate-intl/index.html
Virgil. 1909. *Virgil's Aeneid*. Translated by John Dryden, with Introduction and Notes. New York: P. F. Collier and Son.
Virgil. 1916. *Eclogues. Georgics. Aeneid: Books 1–6*. Translated by H. Rushton Fairclough. Revised by George Patrick Goold. Loeb Classical Library. Cambridge, MA: Harvard University Press.

Part IV

Planetary repair and survival

11 Resisting precarity and planetary dysphoria with *In the Future They Ate from the Finest Porcelain*

Aleks Wansbrough

Introduction: storytelling as a mode of resistance

Planetary precarity looms, with menacing urgency, over contemporary culture. Amid continual profit from the ongoing wrecking of the natural environment, there is a danger of overlooking the relationship between survival, cultural identity, and collective storytelling. Storytelling is one mode of critiquing and resisting a sort of planetary-induced depression, what Emily Apter (2013) poignantly refers to as "planetary dysphoria". Apter's term situates a sense that there is no narrative progress beyond ecological and political collapse. This chapter operates on the assumption that human and political agency is moored in narrative, as a sense of the past and the anticipation of a future is necessary to orient, anchor, and thereby guide, not only ourselves but also emancipatory projects. The video art installation and film, *In the Future They Ate from the Finest Porcelain* by Larissa Sansour and Søren Lind (2016), offers a distinctive approach to the planetary crisis by weaving together themes of cultural, economic, and environmental insecurity with the notion of futurity and storytelling. The work is an experimental film/video that deploys images and tropes from science fiction/speculative fiction.

By examining this film/video, this chapter provides an opportunity to contemplate the interstices of history and precarity. While the film explores specifically the plight of Palestinians, the aesthetic focus on the scorched earth/desert signifies both the literal and symbolic extirpation of a culture. But the work's preponderant imagery of a darkened sky and the barren desert also conjures associations of planetary devastation and the film leans into this interpretation through its use of science fiction tropes. This chapter, therefore, delves into the convergence of planetary and historical narrative critique, and aims to raise questions about potential futures and forms of resistance to precarity in favor of habitability, intimating that we need to explode in a Benjaminian way narratives of despair. The film offers us a way to engage with habitability as it explores both a sense

DOI: 10.4324/9781003383598-16

of melancholia, that I argue chimes with Apter's concept of planetary dysphoria, and portrays a protagonist attempting to lay the grounds for cultural habitability.

Narrative and precarity

The future, not just of human beings but also of other species, is at risk. Precarity is omnipresent in the era of the Anthropocene. Those of us who accept the science of climate change realize that ways of life are under threat from emergency conditions induced by ecological calamities. Yet the people who will suffer most and are already suffering the worst as a consequence of climate change are those who already suffer from economic and political precarity. As the UN Refugee Agency (n.d.) notes, "vulnerable people living in some of the most fragile and conflict-affected countries are often disproportionately affected" (n.p.). Indeed, the Agency elsewhere reports that there is already an annual average of 21.5 million people who have been displaced by sudden weather-related events (2016, n.p.).

With mounting despair over the planet, which has already become an object of mourning, there is no greater cathartic release in contemporary cinema than cosmic death; sating the super-egoic thirst for annihilatory judgment. Rather than turning toward habitability, which in Achille Mbembe's words entail "sharing the earth as opposed to just partitioning it" (YouTube 2022, 19: 56), much of contemporary cultural expression is geared towards apocalyptic narratives. Perhaps, there is an unconscious desire being expressed, namely that we are all guilty consumers and deserve to be punished. If so, this would elevate the prospect of human extinction by imagining it as a kind of divine retribution for human overconsumption. After all, deep ecologist-inflected critics sometimes frame humanity as a virus or cancer on the earth, and books like Alan Weisman's (2007) *The World Without Us* positively reflect on human erasure/extinction, ending with a call to reduce the world's population through a one-child policy. Anti-natalists even hold that human extinction would benefit humanity itself as suffering would also disappear. From this dubious vantage point, the end of the world is also the end of precarity and suffering. It is little wonder then that depictions of planetary annihilation on screen depopulate the world of the downtrodden and hyper-exploited. In films such as Lars von Trier's (2011) *Melancholia*, Adam McKay's (2021) *Don't Look Up* (despite its critique of capitalism), and even Marvel's Russo brother-directed (Russo and Russo 2019) *Avengers: Infinity War*, planetary death obscures climate change's relation to economic and political precarity. In these films, it is as though the poor and exploited were wiped from existence before world-ending calamity struck. Yet for all we know, in reality (in)humanity may not only persist in contrast to fantasies of total extinction, but capitalism may as well, with temperature-controlled environs for the elite becoming the new normal. Meanwhile, the marginalized will continue to endure severe socio-economic conditions, but this time exacerbated by climate catastrophes. Politically, climate refugees will (continue to) be punished, surveilled, and subject to punitive borders as everyday life will become ever-more militarized. Yet recurringly, the climate crisis is metaphorized in spectacularized screenic imaginings: asteroids, comets, or planets colliding with the earth. Such cosmic reckonings can be understood as symptoms of what Fredric Jameson (1994), Slavoj Žižek (1994), and Mark Fisher (2009), have framed as a limit inscribed within capitalism's imaginary. From this perspective "it is easier" and seems even preferable "to imagine the end of the world than the end of capitalism" to borrow the expression of Jameson (1994 xii), Žižek (1994, 1), and Fisher (2009, 1).[1]

Larissa Sansour's and Søren Lind's (2016) work, *In the Future They Ate from the Finest Porcelain* (henceforth referred to as *In the Future*), is distinctive in its exploration of issues of cultural and political precarity along with intimating ecological and planetary catastrophe. The work is both a video installation and a film. As such, it is quite unusual as a work of cinema, lacking overt narrative details, withholding information about the two protagonists, and only featuring symbolic dream images. It stages a narrative and dialogue between anonymous fictional characters, a psychiatrist, and a rebel or "resistance leader", interspersed with dream scenes or representations of the rebel's psyche, often in the form of hallucinatory tableaux. We learn, for instance, that the rebel has been creating porcelain ceramic objects and burying them. Using radiation to obstruct carbon dating, she is trying to create an imagined civilization and past for her people so that future generations will be able to draw strength from such counterfeit antiquities (we learn this wholly through the characters' dialogue). The inclusion of the Keffiyeh pattern on the ceramics strongly intimates Palestinian identity, for it is not only a traditional symbol but has been involved in nationalist struggle, notably by being worn by Yasser Arafat. Dream-laden conjurings of falling bombs that drop ceramics that break against the earth and shatter evoke the annihilatory power of occupation and invasion, as well as the bombings that the Gaza Strip has endured. Indeed, Israel's Operation Cast Lead (2008–2009) was seven years prior to *In the Future*, while Operation Protective Edge (2014) was only two years prior. Not only are people killed, but the cultural links that provide mechanisms of support are eviscerated as are the ties to the landscape on which this cultural heritage breaks.

The landscape appears uninhabitable and uninhabited in the video. Unmoving photographic figures appear against the landscape, but they seem juxtaposed to it, almost as ghosts. They are not people but memories.

The images of ceramics being dropped from bombs also suggest that ceramics can be a weapon, that the shards and imagined relics underscore new narrative possibilities for embodied political struggle against colonial power (Pappe 2006), whereby cultural as well as planetary habitability is at stake. Although it is not stated what society the protagonist inhabits and only through dialogue it is revealed that her culture is forbidden and repressed by some kind of colonial authority, we can surmise that it is an analogue for the repression of Palestinians resisting Israel. In science fiction, after all, images of strange spaceships landing on barren worlds – imagery present in dream tableaux in *In the Future* – indicate colonization. The ceramics tell a story of an imaginary lost identity of the rebel's people, culturally vibrant people before colonization – as an act of imaginative reconstruction.

Sansour's and Lind's film/video poses a challenge to a straightforward narrative encapsulation and, therefore, to resolution and cosmic catharsis. There is little plot and no character development. What we as an audience observe is symbolic imagery and we hear a conversation that includes various philosophical issues of imagining a future. By focusing on dialogue as a mode of storytelling, the work insists on thinking multiple interrelated issues together. It is convenient, after all, to ignore the connections of custom and tradition when thinking about the devastations to ecology and it is also convenient to distance political and ecological precarity, which would leave issues of ecology in the hands of technocrats, government, and corporations. *In the Future* situates the interrelation between culture, land, and violence by showing bombs explode that contain ceramics and also leave the earth scorched, making a connection between the devastation of war or bombing on the earth and the destruction of culture, namely the destruction of porcelain objects with Palestinian patterns. In so doing, the film evidently evokes ties between (settler and planetary) colonization, imperialism, and ecocide. Recognizing these connections, visually the film nevertheless leaves open the possibility of resisting the precarity that results from war and ecocide, even as it aesthetically nods to the vaporization of cultural ties and continuities. For instance, there is no suggestion of direct violence or resistance toward the occupying state in the film. Instead, resistance becomes cultural and symbolic as the rebel protagonist buries cultural objects. Moreover, the film never fully discloses where or when it is set – we do not even know if it is set on earth per se – although all the cultural signifiers indicate that the film concerns Palestine, with Palestinian symbols throughout and the Arabic language spoken. We can conclude that we are in Palestine but a Palestine that is not necessarily entirely of this world. The unspokenness of Palestine renders tangible experiences of loss and alienation, a sense of precarious belonging since viewers are left unsure as to who the rebel is or when this conversation is taking place.

Thus, the lack of details about the rebel's identity underscores the viewer's disorientation and displacement. Indeed, the viewer never sees her exist in a real place, only in dream or fantasy scenes. This withholding of details of her identity, an identity almost wholly virtual and gestured to through symbols, implies that the evisceration of her culture is so complete that it no longer has a name. Indeed, it may recall how the Palestinian flag is periodically restricted in the West Bank (Dalrymple 2002; Najjar 2022).

The film also fuses dream scenes with alien-seeming space crafts, thus indicating further planetary alienation – the earth isn't simply home at this point but another planet. The protagonist is juxtaposed in a fantasy scenario against a scarred and barren landscape as strange alien crafts depart. She is hooded in a uniform that may conjure images from science fiction. It is not even clear who the two voices are, though it appears that one is a rebel and the other a psychiatrist (as revealed in the closing credits). We do not know if the psychiatrist is an interrogator and functionary of the state, or whether she is attempting to help the protagonist deal with personal trauma and abandon her resistance. At several points, the psychiatrist tells her that her dream of a future where her people are free may never come.

In analyzing Sansour's later work, *In Vitro* (2019), Eliza Woods Harrison (2023) uses the term "temporal limbo" to designate the film's un-setting (n.p.). For Harrison, such decontextualized environs are suffused with the trauma of the Nakba (the Arabic word for catastrophe), that is, the expulsion and flight of approximately 750,000 Palestinians in 1948 as part of the creation of the state of Israel. Harrison's observation applies equally to *In the Future* as every image depicting deserts serves as a poignant metaphor for the evacuation of culture and people, for the smashing of identity and means of support. Aesthetically, such hallucinatory imagery functions as a palimpsest of traumas experienced by Palestinians as in the film historical photographs of inhabitants and British colonists are juxtaposed in the form of tableaux. Dadaistically torn from context, such dream compositions of historical and present diremption of cultural habitus often appear to frame the character of the rebel (played by Pooneh Hajimohammadi but voiced by Sansour herself, which further reinforces a sense of split identity). This placelessness provides a sort of silhouette of Palestinian (non)identity. In his analysis of Sansour's futurity, Robert Duggan (2020) cites Edward Said's claim "that today Palestine does not exist, except as a memory or, more importantly, as an idea, a political and human experience, and an act of sustained popular will" (69). For Duggan, Said and Sansour reveal that "without set borders and its own legal and constitutional order, Palestine can only persist in ways heavily mediated by time" (70). As Duggan concludes, "Palestine, for Palestinians, [is] what has been lost, and potentially what will have been created, or perhaps restored [as] moments in time bridged by a contemporary present defined by absence and stasis" (70).

As such, (a)temporality is underscored in terms of narrative, signaling possible narrative progression.

There is an autobiographic, even autographic, quality to the film. Larissa Sansour's work has suffered censorship due to her championing of the Palestinian cause. Her work *Nation Estate* (2011) was excluded from the Elysée art prize competition due to the sponsor, Lacoste's insistence. The Musée de l'Elysée decided to cancel the competition amid the controversy. Sansour alleges that the museum told her that Lacoste considered her work "too pro-Palestinian" (Milmo 2011, n. p.). The protagonist of *In the Future* is described as a "narrative terrorist", which perhaps could function as a description of Sansour. The rebel in the film is voiced by Sansour, thus conflating her to a degree with the protagonist and suggesting an affinity between them, but the fact that the rebel is performed by a different actor affords critical reflection, engendering spectatorial uncertainty. Being Palestinian, born in Bethlehem but residing in London, Sansour presents work that navigates intergenerational trauma, memory and postmemory amid the evisceration of place and identity (AlAmmar 2019), in which the disembodied voice of the protagonist accentuates the precarity of Palestinian presence. The notion of precarity is foregrounded through the revelation that the rebel lost her sister who was killed by the occupiers at the age of nine. As the rebel confides, "they mistook her for something else. They didn't know what they killed". That such tragedy is not an anomaly is explained by the rebel: "After a certain point, death is no longer about the single life lost. It's not even personal. It's what we are as a whole that qualifies us as targets" (Sansour and Lind 2016).

Although the film reveals that precarity is what Judith Butler (2015) calls a "politically induced condition in which certain populations suffer from failing social and economic networks of support more than others, and become differentially exposed to injury, violence, and death", it also assumes a planetary dimension (33). Gil Z. Hochberg (2018) considers *In the Future* to be part of a contemporary turn toward dystopic science fiction, speculating that "the ease with which dark futurism and dystopic imagination circulate these days may indicate the genre's potential for raising awareness of our planetary existence and shared destiny" (35). The use of science fiction elements necessarily calls attention to planetary issues such as the future colonization of distant worlds, and also the idea of barren planets, planets rendered inhospitable that may foretell earth's future. Thus, the work accentuates climate precarity. Indeed, the West Bank and Gaza are suffering from climate change (Climate Change Knowledge Portal n.d.) with the International Red Cross deeming Gaza to be "on the frontlines of climate change" with rising tides and extreme weather conditions (International Committee of the Red Cross 2022). Moreover, Gaza has figured in recent political discourses as a metaphor for a future subjected to the extremes of planetary and military violence. Recently, Columbian President Gustavo Petro called the condition of Gaza a "rehearsal of the future" (Fadul 2023). Moreover, the tableau imagery from *In the Future* features images of refugee camps and tents with flying crafts above. While such images depict Palestinian displacement in the wake of the Nakba, they may seem to prophesy climate refugees if we do indeed understand Palestine to be a "rehearsal of the future".

The recurring images of strange flying crafts in *In the Future* makes explicit the idea of planetary conquest and extraction, of the destruction and harvesting of resources as seen in science fiction and recent historical reality. In such narratives, planetary conquest maps capitalist extractive practices and colonial crimes. Indeed, interstellar conquest, invasion, and resource extraction have always been metaphorized through science fiction, with science fiction verging on becoming an imaginative form of reportage. Beginning with H.G. Wells's ([1897] 2011) *War of the Worlds*, colonizers have often been rendered aliens, guilty of both genocide and ecocide:

> [B]efore we judge of them [the Martians] too harshly we must remember what ruthless and utter destruction our own species has wrought, not only upon animals, such as the vanished bison and the dodo, but upon its inferior races. The Tasmanians, in spite of their human likeness, were entirely swept out of existence in a war of extermination waged by European immigrants, in the space of fifty years. Are we such apostles of mercy as to complain if the Martians warred in the same spirit?
> (Wells [1897] 2011, 7)

Wells's passage tethers the extirpation of other species to that of groups of human beings. Although using the racist language of species, the passage nevertheless narrativizes the murder of both people and animals. Such themes continue to be explored in science fiction today from the novels of Kim Stanley Robinson or James Cameron's (2009) *Avatar*. As these same themes recur through science fiction narratives, there is perhaps no more fitting vehicle to reveal the entanglement of planetary and cultural precarity than science fiction.

In the Future leans into this Wellsian conceit of equating planetary conquest with alien technology. Cleverly, the film/video never shows who is doing the occupying, though. Thus, the film evades the risk of an anti-Semitic portrayal of Jewish Israelis as alien beings. We see images of futuristic ships leaving a now-barren location as though departing from the planet, as well as images of British colonialists standing against this seemingly uninhabited locale, as if ghostly photographs from the past gesturing also to Palestine's history. Such images remind viewers of the British

Mandate for Palestine, and the Balfour Declaration, which aimed to carve up the land by creating a homeland for Jewish Zionists. Indeed, *In the Future* adopts a planetary aesthetic, conjuring a sense of apocalypse and a dying planet through science fiction generic overtures in featuring mesmeric and phantasmal scenes of dark sands that appear scorched and skies at sunset that bear a blazing hue. In fact, Sansour's next film – namely *In Vitro* (2019) – would explore the aftermath of an ecological tragedy. Her previous work, *Nation Estate* (2011), explores the idea of an apartment block of Palestinians with Palestinian shrubs inside the building. As a middle installment in this series of works, *In the Future* can similarly be understood as exploring ecological dynamics. After all, the earth or ground features prominently in the dream imagery.

In *In the Future*, the earth becomes a type of character infused with contradictory and interwoven meanings, functioning as both the site of lost civilization and buried trauma. In a dream scene, the rebel stares at a hole in the soil that she has dug, with the camera even adopting an angle of vision from the hole of the rebel, rendering the terrestrial chasm a kind of subject or perspective. If subjectivity is an abyss, as some thinkers have maintained, if it depends on a gap in causality, then the earth itself is rendered a kind of subject. The hole is mysterious and deep and comes to shape the land and also the rebel's struggle. As the rebel stares into the abyss, the absence structures both her subjectivity and the land itself as the sense of historical being and of personal trauma and grief abyssally converge. It is unclear, then, whether the emptiness within the land is generative or whether it reflects only deep mourning, a pit of sorrow. This massive hole in the ground could be a tomb, an archaeological site, the burial of her sister, a well or spring, or the crater left by a bomb blast. It is an earthen wound, and as a cavity it calls to mind the Anthropocene, commonly defined as the period of geological time where humans have made a dominant impact upon the earth. Indeed, the Anthropocene is sometimes dated to thermonuclear bomb tests (Subcommission on Quaternary Stratigraphy 2019, n.p.).

Therefore, while the scenario is specific to Palestine, much of the discussion between the rebel and the psychiatrist is more general and so points to a larger political and environmental crisis. For instance, at one point, the psychiatrist states: "The apocalypse never is that single cataclysmic event. It sneaks up on you", which may recall the durational quality of the slow violence (Nixon 2011) of climate change. The rebel seems to make the comparison herself between Palestine and global cataclysm when she considers: "this place was always a barely functioning dystopia deeper into the apocalypse, an accelerated microcosm. It all disappears little by little" (Sansour and Lind 2016). As a microcosm, Palestine becomes not merely a contested and precarious territory and the Palestinians an immiserated people, but also prophetic of a near future, with territory and resources becoming scarce. Read from this angle, the psychiatrist's statement that "the descendants claiming the land of their fictional ancestors, they will never exist", obviously referring to Palestinians, also resonates with the possibility that humans will become extinct amid planetary collapse and climate feedback loops. The imagery similarly gestures to the planetary, with its emphasis on land and soil.

Cannily, the film equates memory with the earth itself. According to Walter Benjamin ([1932] 2005), earth is the site of memory, whereby "it is the medium of that which is experienced, just as the earth is the medium in which ancient cities lie buried" (576). There is perhaps an unintended ecological dimension to Benjamin's pronouncement. If memory is like the earth and both are mediums rather than tools, then the earth's instrumentalization is an affront to its ability to mediate all things. Indeed, the film enacts a struggle between the earth as a site of instrumentalism, war, and territory versus the earth as a refuge of meaning, as a site of growth and memory. To search for memories, argues Benjamin, is to dig in the ground. Rather than yielding an unmediated whole, memory, like the earth is to be scattered, as one "turn[s] it over as one turns over soil" (576). The earth is positioned as a manifestation of the psyche and vice versa in both Benjamin and Sansour, opening up a discussion of how earth and memory can be tied together.

Dysphoria and narrative closure

By tethering cultural memory, earth and psyche together, the film renders the earth as a zone of contested meanings, but also as a possible tomb of despair. From a psychoanalytic vantage point, the hole could at once symbolize the womb – the earth being synonymous with the mother, and nothingness with undifferentiated being (one is one with the mother in the womb), but at the same time, the chasm invokes separation and estrangement, as the earth is no longer a totality, no longer whole, and thus the human political subject, like the earth itself, is marked by a gap. The earth is sorrowful and rendered barren, inhospitable, and uninhabitable. In this way, the film has a curious relationship to Emily Apter's (2013) concept of "planetary dysphoria". Planetary dysphoria, according to Apter, is present in philosophy such as speculative materialism, contemporary art, and cinema, pervading contemporary culture (138).

As with romantic lamentation, planetary dysphoria denotes a melancholy disposition. Using Freudian terminology, Apter associates planetary dysphoria with notions of despair about the planet and ecology, a sense of powerlessness and guilt. She notes that:

> For Freud, there was in fact a planetary dimension to melancholia: "In mourning it is the world which has become poor and empty; in

> melancholia it is the ego itself." Melancholia extends the sense of the world as "poor and empty" to the ego.
>
> (2013, 139; quoting Freud [1917] 1995, 244, 246)

Further quoting Freud's assessment of melancholia as "a delusional expectation of punishment" (244), Apter argues that the "internalization of anger against self" is ultimately "masochistic and suicidal at its most extreme" (139).

Apter's formulation of planetary dysphoria thereby helps to accentuate pathological and hyperbolic responses to planetary crisis. According to Apter, planetary dysphoria finds its apotheosis in "a thanotropic projection of how a planet dies" (2013, 140). For her, planetary dysphoria is a "'depressive position' suffusing every aspect of economic, social and terrestrial life" (139). It manifests as a sort of self-torture "captur[ing] the geopsychoanalytic state of the world at its most depressed and *unruhig*, awaiting the triumphant revenge of acid, oil and dust" (140). Apter presents us with a pathological, death drive concept of destruction that reflects contemporary concerns. The fixation on becoming cosmic dust is a manifestation of a sense that our agency is thwarted, no change is possible, and all that remains is punishment for our collective guilt. Death is easier to imagine than alternative social and political structures that cause precarity. Rather than overcoming precarity and building networks of resistance, planetary dysphoria surrenders to apocalypticism and a pseudo-karmic judgment. One may extend Apter's views, that with planetary dysphoria the earth is no longer conceived of as a home or locus, but potentially as a burden and threat, an object that has been violated but also contaminated by humanity that itself must be purged from the universe.

On the face of it, there is some proximity between *In the Future* and planetary despair. After all, the character of the rebel bears some resemblance to the classical embodiment of melancholy par excellence, namely Antigone, who is effectively buried alive in a tomb as punishment for burying her brother. The rebel, too, buries objects – though not to mourn but to conjure a future. Nevertheless, a mood of latent, teetering despair enshrouds the images as we witness barren desertscapes and a chasm in the ground. The cinematographic fixations on dust and spaceships in Sansour's work could indeed be "a thanotropic projection" of planetary death. The emphasis on tableaux and slow motion, leaves one with a feeling of stasis or else intimate Rob Nixon's (2011) "slow violence" (whereby planetary violence can seem imperceptible due to the durational inducement and occurrence of such devastation), replete with inexorable dread. Indeed, Ari Akkermans (2016), writing for *HyperAllergic*, observes that

> the digital film is set in a dystopian territory without a future, or at the very end of historical time. The master narrative of the end-of-times is

> not an event but a condition: Disaster becomes not sheer bad luck, but a fixed lens through which history is narrated.
>
> (n.p.)

Yet, despite the melancholic atmosphere, the film is sufficiently philosophical to forego the pathology of a sacrificial planet as its lack of resolution does not lead one to despair alone but may invite constructive questioning. The film is also sufficiently resistant to planetary dysphoria as the planet is not wiped out in a cataclysmic spectacle. This in part explains why Sansour has argued that "glossy sci-fi is preferable to us always being the subjects of documentaries, which is something I think is hurting us, more than anything" (Sansour 2016, n.p.). Documentaries risk becoming nothing more than documents in a catalogue of subjugation, as an unintentional enforcing and re-enforcing of despair. Sansour elaborates that

> we always project a state; but at the same time, we always talk about the Nakba (lit. 'Catastrophe'), so there is always this dialogue between nostalgia and ambitions for the future. The present is kind of lost in documentaries; it lives in this state of limbo, it isn't real. That's why I think sci-fi works really well.
>
> (2016, n.p.)

Dysphoria entails a sense of backwardness, as through planetary collapse has already happened. The future is gone and so we have lost the present. Indeed, such spaceships may evoke the idea that at the end of the world, only aliens will uncover the buried artifacts. In this way, the work allows us to critically consider what the rebel protagonist refers to in her dialogue as a "bereavement both material and aesthetic" (Sansour and Lind 2016). In Freudian terminology the film is more healthily concerned with mourning than with melancholia (Freud [1917] 1995, 243–244) – which is to say it does not project defeat and cosmic justice, but turns toward real loss. As such, the film raises the possibility of human justice and human beings creating new narratives by which to survive colonization, imperialism, and ecocide. It gestures to the possibility of surviving precarity.

From one angle then, the film/video is what it explores, a future-oriented work of mythic invention. Yet, it leaves the viewer in a state of uncertainty because it is not straightforwardly nationalistic. As Sansour explains, "the film posits a philosophical equation rather than really creating a nation" (2016, n.p.). As political subjects we find our sense of continuity dependent on narrative projection and forward momentum and the film/video addresses this issue. At first, it may seem to indicate a tendency toward regression – as identities and peoples are shattered, the most one can muster in resistance is to imagine an alternative past that never existed. But precisely here we discern an admission that identity itself is constructed.

Dialectically, far from a retreat to an idealized past for present sustenance, the past is the creation of the present in order to conceive of a future that may yet be. Through the idea of counterfeit relics, there is an underscoring that history can be rewritten and transformed. Curiously, the future depends on reimaginings of the fragments of the past. In this respect, there is something messianic about the rupture, something Benjaminian.

Messianic rupture and narrative reconfiguration

In his *Theses on History*, Walter Benjamin ([1940] 1968) advocates the concept of messianic rupture as he navigates different forms of Marxism. For Benjamin, "the class struggle, which is always present to a historian influenced by Marx, is a fight for the crude and material things without which no refined and spiritual things could exist" (254). While agreeing that material things are necessary for higher forms of knowledge, Benjamin nevertheless seems to affirm those higher aspects of art, culture, and theology. He asserts that ultimately Marxism, to be truly self-conscious, must understand its relation to theology. The idea of a deterministic worldview whereby history has only one path to socialism mirrored a sort of Christian predestination. Behind this mechanistic formulation of history was, claims Benjamin, "a little hunchback", namely the hunchback of theology (([1940] 1968), 253). Instead, Benjamin advocates a different messianic theological framing of history, one that would explode the past and borrow from its fragments. Benjamin asserts that

> nothing that has ever happened should be regarded as lost for history. To be sure, only a redeemed mankind receives the fullness of its past – which is to say, only for a redeemed mankind has its past become citable in all its moments.
>
> (254)

While Benjamin's analysis is much contested, I take him to be saying that rather than approaching history as a narrative – the frozen past – we should instead take history to be an opportunity for future reconstitution, in that "every image of the past that is not recognized by the present as one of its own concerns threatens to disappear irretrievably" (254).

Benjamin, that is, focuses on memory not as a unified whole, but rather as scattered trinkets. Exploring the theme of memory and history as shattered, Benjamin argues that history is a continual struggle between oppressor and oppressed. From one perspective, he argues, this class struggle has always favored the victor, but from another, this struggle is far from over and so a victory by the oppressed will necessarily alter past struggles, transforming them into a triumph. Benjamin insists that "Articulating the past historically does not mean recognizing it 'the way it really was'. It means to seize hold a memory as it flashes up in a moment of danger" ([1940] 1968, 255).

The messianic quality has a curious relationship to narrative and catastrophe. By definition, the messianic ruptures narrative but it also implies narrative, as a sort of occurrence that changes and reforms narrative. The film itself plays with the idea of the messianic, of the now as a way of creating a future, a turning point. Theological imagery is played with throughout. Indeed, the rebel describes having dreams where she becomes her people's Shroud of Turin. This would seem to underscore the messianic quality of revolt and the idea that revolutionary change requires a fidelity verging on religiosity – after all, Christ's death was a loss that retroactively became a victory. The messianic is mythical but it speaks to a yearning to interrupt history and reinterpret it. In Sansour's film then, there is an overt need to borrow from even the colonizer and settler, to explode their power through their archaeological methods. The rebel states poignantly:

> It was already a frontline. Our rulers built a nation on archaeology. It's no longer about history. It's an epistemology, a tool for shaping a national imagination. Projecting a state into the past supports the idea of historical entitlement […]. In its most perverted form, archaeology galvanizes public sentiment, confirms myths of the past, and defends them against scrutiny.
>
> (Sansour and Lind 2016)

What we discern in this quote is not a defense of nationalism per se, but an acceptance of fragmentation and the need for strategy and contingency, to explode the certainties of history and historical defeat.

This recalls Benjamin's position and the vitality of his arguments when used to combat the very types of narratives that have become popular amid planetary dysphoria: narratives both in terms of fiction – such as end-of-the-world narratives which foreclose upon anything other than planetary destruction – and in terms of a cultural understanding of the present in relation to the past and the future. Rather than assume that a narrative must be resolved a certain way, and that fate is written in the stars, Benjamin attempts to defy and liberate us from fatalistic convictions and conventions. Although Benjamin was arguing within Marxism against a deterministic form of Marxism, his arguments remain relevant in considering the politics of history, narrative, and resisting a sense of foreclosed futures. Even if Marxism has withered away as a political force, Benjamin has newfound relevance. Amid planetary precarity, there has been the predominance of disempowering narratives, namely that only elites can rescue us, that tech billionaires like Elon Musk may yet provide us the means of escaping the earth's ties. Effectively, we have lost a sense of historical – that is, collective agency. Protests against the Iraq war did not prevent the Iraq war. Protests against Wallstreet have not dented their profits. There is a sense in which we have become in Timothy Morton's and Dominic Boyer's (2021) words "hyposubjects", aware of our precarity upon the earth, dwarfed by hyper-objects (13–14), and consumed by planetary dysphoria, by a sense of inevitability.

There are undoubtedly a variety of causes for this loss of collective agency. One may, for instance, point to the internet and how the digital breaks subjects down to numbers in what Gilles Deleuze (1992) considers in terms of "dividualism" (4). For Deleuze, the computerized societies of control replace subjects with fractions, manipulating people not as disciplined subjects but as reconstitutable metrics, engineering a form of "free-floating control" (4). Then there are the economic forces and the turn toward neoliberalism, which, as David Harvey (2005) has persuasively demonstrated, has depleted human agency (191). Individuals have more individual agency when they can make a difference within a cause or collective, and can rely on supportive structures, social organizations with recourse to mobilization. With the smashing of union power, the ability of individuals to organize and engender change has dissipated. Guy Standing (2011) has popularized the phrase the precariat to locate "a new dangerous class", easily co-opted by populist movements that have emerged from the push toward employer flexibility, work automation, and the abandonment of Keynesian economics.

Yet, the film/video and Benjamin's theories afford a different way of thinking about these events. It is precisely amid such despair that both Benjamin and Sansour offer the potential to think both in narrative terms, and to explode narrative. Rather than seeing this crisis as the culmination of various tendencies, both Sansour and Benjamin posit that we have never been without crisis. The crisis itself juxtaposes various struggles that, rather than being conceived of as over, can be understood as ongoing. The Anthropocene is not merely a time of spectacular and frightening climate occurrences, but also a time of loss, disappearance, and silence. As some recent climate protestors have said, there is no art on a dead planet. Cultural narratives and precarity are entwined. Ella Harris and Mel Nowicki (2018) contend:

> there remains limited attention regarding how precarity has been *culturally*, as well as socially and economically, entrenched. This is despite the fact that, as we argue, the reproduction of precarious conditions is now, perhaps most significantly, a cultural reproduction, operating through imaginaries.
>
> (388)

So too, with resistance to precarity, as culture, history, and narrative pose impediments to extractive capitalism, (neo)colonialism, and ecocide, offering an expectation of the future. Cultural evisceration, therefore, whether by the state of Israel, Putin's Russia, the People's Republic of China's persecution of the Uyghur population, or western capitalist nations, is a necessary precondition to the carving up of land and resources. Precarity entails the loss of stability and of world, of the support processes necessary for resistance. It speaks to the global.

Conclusion: narrative ambiguity as narrative contestation

While many studies of precarity concern neoliberal capitalism, Sansour and Lind's work returns to a precarity that predates neoliberalism, one that is related to Indigenous cultures dispossessed amid the forceful expropriation of land. Yet, according to various cultural theorists (Apter, Jameson, Fisher, and Žižek), cultural portrayals of planetary annihilation and planetary precarity stem from neoliberal capitalism, tethering contemporary science fiction to planetary doom. Sansour and Lind explore themes of devastation, both planetary, human, and cultural in a way that conjures images prior to neoliberalism (the Nakba) and that may continue well beyond the present, offering an image of resistance to planetary dysphoria. The film/video does not conclude but leaves the future open. In a dream sequence, we see the rebel and her deceased sister hold hands briefly before the sister leaves. The dual possibility of being reunited and of a final loss haunts the close of the film. Can the dead be redeemed by the future? The film offers no answers, easy or otherwise. But it does accentuate the way that narrative and history are not static but forever subject to contestation. The ending may be ambiguous, but such ambiguity affords a historical openness to possibilities of resistance rather than submersion in despair. As we face precarity as a species, and as various individuals and populations face a variety of precarities, the dominance of narrative and its explodability, as a means of resistance should come to the fore. Thus, the film explores issues of narrative and agency, but does not cling to an imagined home – rather it concerns survivability and the need to construct a more habitable world. Indeed, the faked relics exist not to gesture to some imagined totality of a complete history but rather to challenge the very idea of history, suggesting that even an envisioning of a different type of future is a radical gesture – an opening for an almost Benjaminian messianic understanding of history rather than an enclosing, suffocating mood of planetary dysphoria. For Walter Benjamin and the rebel of *In the Future*, it is by reimagining and reinterpreting the past that new futures can be conceived.[2]

Notes

1 The phrase has its origin in Jameson's (1994) critique of environmental devastation, namely, "it seems to be easier for us today to imagine the thoroughgoing deterioration of the earth and of nature than the breakdown of late capitalism" (1). The sentence was refined, modified, and popularized by Žižek and Fisher into the more recognizable expression that it is easier to imagine the end of the world than the end of capitalism.

2 This chapter was largely written prior to the atrocities of October 7, 2023, and the Israeli "response", which South Africa and human rights experts accuse of genocide. Further, the planetary and ecocidal dimension of what is happening to Gaza, including the intentional destruction of farmland, necessitates collective agency and Palestinian stories to be told across the world.

References

Akkermans, Ari. 2016. "A Sci-fi Film Imagines a Dystopian Future for Palestine." *HyperAllergic*. April 26. Accessed February 20, 2023. https://hyperallergic.com/293728/a-sci-fi-film-imagines-a-dystopian-future-for-palestine/

AlAmmar, Layla. 2019. "Palestinian Postmemory: Melancholia and the Absent Subject in Larissa Sansour's *In Vitro*, Saleem Haddad's 'Song of the Birds', and Adania Shibli's *Touch*." *Journal of Literature and Trauma Studies* 8 (1): 1–23.

Apter, Emily. 2013. "Planetary Dysphoria." *Third Text* 27 (1): 131–140.

Benjamin, Walter. [1932] 2005. "Excavation and Memory." In *Selected Writings, Volume 2: 1927–1934*, translated by Rodney Livingstone and edited by Michael W. Jennings, Howard Eiland, and Gary Smith. Cambridge, MA: Belknap Press of Harvard University Press.

Benjamin, Walter. [1940] 1968. "Theses on the Philosophy of History." In *Illuminations*. Translated by Harry Zohn, 253–264. New York: Schocken Books.

Butler, Judith. 2015. *Notes toward a Performative Theory of Assembly*. Cambridge, MA: Harvard University Press.

Cameron, James, dir. 2009. *Avatar*. United States. Twentieth Century Fox.

Climate Change Knowledge Portal. n.d. "West Bank and Gaza." Accessed May 27, 2024. https://climateknowledgeportal.worldbank.org/country/west-bank-and-gaza/extremes

Dalrymple, William. 2002. "A culture under fire." *The Guardian*. October 2. Accessed February 10, 2024. https://www.theguardian.com/artanddesign/2002/oct/02/art.artsfeatures

Deleuze, Gilles. 1992. "Postscript on the Societies of Control." *Cultural Theory: An Anthology*59: 3–7.

Duggan, Robert. 2020. "Larissa Sansour and the Palestinian Ruins of the Future." *Journal for Cultural Research* 24 (1): 69–83.

Fadul, Laura Gamba. 2023. "'What We See in Gaza Is the Rehearsal of the Future': Colombian President at UN Climate Summit." *Anadolu Ajansı*. December 2. Accessed May 27, 2024. https://www.aa.com.tr/en/politics/-what-we-see-in-gaza-is-the-rehearsal-of-the-future-colombian-president-at-un-climate-summit/3070873

Fisher, Mark. 2009. *Capitalist Realism: Is There No Alternative*? Winchester: Zero Books.

Freud, Sigmund. [1917] 1995. "Mourning and Melancholia." In *On the History of the Psycho-Analytic Movement and Other Work,* translated by James Strachey, 243–258. London: Hogarth Press.

Harris, Ella, and Mel Nowicki. 2018. "Cultural Geographies of Precarity." *Cultural Geographies* 25 (3): 387–391.

Harrison, Eliza Woods. 2023. "Larissa Sansour's 'In Vitro' Images a Post-Apocalyptic Palestine." *Frieze.* September 21. Accessed February 1, 2023. https://www.frieze.com/article/larissa-sansours-vitro-imagines-post-apocalyptic-palestine

Harvey, David. 2005. *A Short History of Neoliberalism.* Oxford: Oxford University Press.

Hochberg, Gil Z. 2018. "'Jerusalem, We Have a Problem': Larissa Sansour's Sci-Fi Trilogy and the Impetus of Dystopic Imagination." *The Arab Studies Journal* 26 (1) (Spring), 34–57.

International Committee Red Cross. 2022. "Gaza: on the Frontlines of Climate Change." Accessed May 27, 2024. https://www.icrc.org/en/document/gaza-frontlines-climate-change

Jameson, Fredric. 1994. *Seeds of Time.* New York: Columbia University Press.

Mbembe, Achille. 2022. "Climate, Sustainability & Inequality Seminar: Notes on Planetary Habitability by Prof Achille Mbembe." *Wits University Official.* November 22. Accessed February 20, 2023. https://www.youtube.com/watch?v=d1AKvMN5ock

McKay, Adam, dir. 2021. *Don't Look Up.* Netflix.

Milmo, Cahal. 2011. "Lacoste Accused of Attempting to Censor 'Too Pro Palestinian' art." *The Independent.* December 21. Accessed February 1, 2023. https://www.independent.co.uk/arts-entertainment/art/news/lacoste-accused-of-attempting-to-censor-too-propalestinian-art-6280082.html

Morton, Timothy, and Dominic Boyer. 2021. *Hyposubjects: On Becoming Human.* London: Open Humanities Press.

Najjar, Farah. 2022. "The Palestinian Flag: A Target for 'Erasure' by Israeli Forces." *Aljazeera.* May 13. Accessed February 8, 2024. https://www.aljazeera.com/news/2022/5/13/why-is-israel-afraid-of-the-palestinian-flag

Nixon, Rob. 2011. *Slow Violence and the Environmentalism of the Poor.* Cambridge: Harvard University Press.

Pappe, Ilan. 2006. *The Ethnic Cleansing of Palestine.* Oxford: One World.

Russo, Anthony, and Joe Russo, dir. 2019. *Avengers: Endgame.* California: Marvel Studios.

Sansour, Larissa. 2016. "Unbreakable: Interview with Larissa Sansour." With Abdellatif R. Abdeliawad. Reproduced on Mecfilm.com (n.d.) Originally published in *Reorient: Middle Eastern Arts and Culture Magazine.* February 16. https://mecfilm.com/fileadmin/user_upload/documents/filmhefte/en/InTheFuture_engl.pdf

Sansour, Larissa, and Søren Lind, dir. 2016. *In the Future They Ate from the Finest Porcelain.* Berlin: Mec Film.

Sansour, Larissa, and Søren Lind, dir. 2019. *In Vitro.* Berlin: Mec Film.

Standing, Guy. 2011. *The Precariat: The New Dangerous Class*. London: Bloomsbury Academic.
Subcommission on Quaternary Stratigraphy. 2019. "Working Group on the 'Anthropocene'." May 21. Accessed February 20, 2023. http://quaternary.stratigraphy.org/working-groups/anthropocene/
Trier, Lars von, dir. 2011. *Melancholia*. Denmark: Zentropa Entertainment.
UNHCR. 2016. "Frequently Asked Questions on Climate Change and Disaster Displacement." Accessed February 20, 2023. https://www.unhcr.org/uk/news/latest/2016/11/581f52dc4/frequently-asked-questions-climate-change-disaster-displacement.html
UNHCR. n.d. "Climate Change and Disaster Displacement." Accessed February 20, 2023. https://www.unhcr.org/en-au/climate-change-and-disasters.html
Weisman, Alan. 2007. *The World Without Us*. London: Picador.
Wells, H. G. [1897] 2011. "The War of the Worlds." Reproduced by Planet PDF. Accessed February 20, 2023. https://www.planetpublish.com/free-ebooks/401/the-war-of-the-worlds/
Žižek, Slavoj. 1994. *Mapping Ideology*. Verso: London.

12 The precarious case of the zero-waste solution to the planetary problem

Kanak Yadav

Introduction: precarious lives and consumable objects

> Objects are neither a flora nor a fauna. And yet they do indeed give the impression of a proliferating vegetation, a jungle in which the new wild man of modern times has difficulty recovering the reflexes of civilization.
>
> (Baudrillard 1998, 25)

French philosopher Jean Baudrillard's (1998) comparison of objects with wilderness and "vegetation" to underline their ubiquity appears paradoxical in the context of climate change and its detrimental impact on the planet. One truism that has emerged from the concept of the Anthropocene, as coined by Paul Crutzen (2006), is how the natural environment is dynamic, constantly changing in ways that are unalterable due to the decisive role that "mankind" has come to play in shaping the "earth and atmosphere" (16). Therefore, even semantically, the presumption of the natural environment as a given seems problematic when nature itself is threatened by human actions. Nonetheless, Baudrillard's comparison between man-made objects and natural vegetation denotes the power that the human too has acquired to easily overpower nature under late capitalism. Thus, Baudrillard's imagery of objects growing like a "proliferating vegetation" brings out the irony of modern consumptive practices since it evokes the visual image of man-made, non-biodegradable products like plastic as drastically "colonizing" the planet.

This chapter opens with a discussion of the omnipresence of consumable objects not only because they pose an obvious risk to environmental sustainability, but also because of the rise of "zero-waste" cultural politics, a response to the problem of environmental precarity, as one that refuses capitalist consumption. Our contemporary moment is marked by precarious living, whether as Guy Standing's (2011) faceless, invisible "precariat" that emerges as a "fragmented global class structure" (7) or as increasing political and ecological refugees, and the impending threat of irreversible

DOI: 10.4324/9781003383598-17

climatic change. Lives are thus marked by a precarious existence organized along the lines of racial, gendered, and social inequalities that have only been further strengthened by the Covid-19 pandemic. Apart from showing the agential role of humans and their impact on the environment, the pandemic also reignited social anxieties pertaining to racial and cultural minorities. The concept of "ecoprecarity", as developed by Pramod K. Nayar (2019), draws attention to the entangled lives of humans, environmental shifts, and non-human living beings, emphasizing that planetary changes cannot be viewed as a "singularity" (9). In Nayar's words, "ecoprecarity is at once about the precarious lives *humans* lead in the event of ecological disaster [...] and also about the environment itself which is rendered precarious due to human intervention in the Anthropocene" (7; original emphasis). Hence, environmental precarity cannot be viewed as an isolated condition distinct from other risky situations and political instabilities that humans have to deal with.

Amidst such a context of environmental politics, the emergence of "zero-waste culture" as a lifestyle-based choice that involves the educated middle-class and its consumption practices is a cultural phenomenon that connects mainstream capitalism to the planetary crisis. This chapter is interested in the cultural practice of zero-waste, its digital aesthetics and the individual-centric solution it offers to questions posed by the risk of environmental sustainability. Although the nomenclature of "zero-waste" has been part of the climate change vocabulary, it has attained widespread popularity through digital media's documentation of zero-waste living and culture. For this reason, this chapter will focus on comparing two culturally different zero-waste blog-cum-e-commerce websites – from the Global North is American environmentalist Lauren Singer's blog "trashisfortossers.com" and from the Global South is Indian environmentalist Sahar Mansoor's website titled "barenecessities.in" – to analyze the politics of zero-waste digital environmentalism and its self-centric, neoliberal solutions to environmental precarity.

The rise of zero-waste

The term "zero-waste" has had different meanings across time frames and cultures. First used by Paul Palmer in 1974, "zero-waste" in his terms implied reusing products instead of discarding them. Palmer (n.d.) explains that by naming his company "Zero Waste Systems Inc.", he introduced the nomenclature which has since attained widespread currency but still hasn't resulted in effective action (1). According to Christof Mauch (2016), Palmer's conception of "zero-waste" was mainly utility-driven, aimed at reducing chemical waste by using products to their full efficiency (6). The way the concept of "zero-waste" is understood today is not in terms of economic efficiency attained through judicious usage of resources, but rather in terms of the environmental impact of reduced or zero wastage. In this context, Palmer argues that "true Zero Waste" should be directed towards "redesigning" products so that they are better suited for multiple uses and can address the fundamental problem of waste generation itself (n.d., 2). While contemporary usage of the term focuses mainly on waste disposal and its effective management, Palmer argues that "zero-waste" should instead be about "eliminat[ing] garbage" by regulating production (n.d., 3). The concept of "zero-waste", therefore, includes many different thought processes instead of a singular framework.

Despite the prevalence of contradictory perspectives in defining "zero-waste" philosophy, it has gained momentum because its call for sustainability brings together the household, society, and governing agencies (local/national/international) to take responsibility for battling climate change. According to Paul Connett (2013), the philosophy of "zero-waste" acquired cultural resonance around 1995 in "places like California and Italy", when the ideology of reduced consumption and reusing/repurposing waste was gaining primacy (n.p.). The digital culture of "zero-waste" should, therefore, be understood as a post-millennium movement following an environmental philosophy that spreads awareness about the problem of waste through the power of digital technology. Taking advantage of the widespread reach of the digital, zero-waste websites popularize sustainability and make zero-waste living into a global movement.

In becoming a global phenomenon through digital cultures that transcend national boundaries, the culture of zero-waste has created a global, transnational community. It started out as a lifestyle choice popularized by individual bloggers in the Global North as they documented their transition towards an entirely unique lifestyle that produced little to no trash. The rise of women bloggers like Bea Johnson, Kathryn Kellogg, and Lauren Singer has mainly contributed to establishing this cultural trend (Anon 2023, n.p.). However, their cultural visibility is primarily associated with their digital personalities, created through carefully chosen images that represent their environmental vision. The iconoclastic image of the self-styled environmentalist holding a mason jar filled with years of trash has become representative of not just the politics and ecological impact of the movement but also its visual aesthetic of minimalism, epitomized by remains contained in a glass bottle. Whether it is Bea Johnson or Lauren Singer, the quintessential image of presenting a trash-filled glass jar has acquired a symbolic value in defining the philosophy of zero-waste culture.

Zero-waste politics, in terms of an environmental movement, has its origins in the Global North; however, it would be a mistake to view it as an entirely new concept without any historical precedent. The ethos of less consumption, minimalism, and reduced wastage stems from an age-old wisdom rooted in non-western cultures and religious practices. The aim of using resources sensibly, avoiding wastage, and refraining from over-indulgence, be it body or mind, is central to many Asian cultures. For example, in Hinduism, Jainism, and Buddhist religions, the path towards spiritual enlightenment is realized through the renunciation of material attachments and by following the path of *ahimsa* or non-violence. Even Gandhian philosophy originates from his belief in Hinduism and its cultural superiority over western capitalism, morality, and modernity (Gandhi [1909] 1997, 37–38). Gandhi's ([1909] 1997) *Hind Swaraj* became a prominent example of his political thought being shaped by his spiritual belief in the supremacy of Indian civilization in that "civilization" is perceived as a "mode of conduct which points out to man the path of duty" (67). Gandhi observes that the path of true civilization is that laid down by the ancestors, who not only "set a limit to our indulgences" but also "dissuaded us from luxuries and pleasures" (68). Hence, the philosophy of minimalism and its spiritual virtue forms an integral part of Gandhian anti-colonial thought. Zero-waste cultures, whether located in the Global South or the Global North, renew and remodel this principle of simple, harmonious living into a globalized, neoliberal twin called the "zero-waste" lifestyle, one that addresses the crisis of environmental degradation by interrogating capitalist consumption and promoting ecological capitalism, carefully fashioned through environment-centric politics and aesthetics.

Zero-waste politics can thus be interpreted as the environmentalism of the current neoliberal age as it unites traditional wisdom with the modern, digital world to battle the planetary crisis. The propagators of zero-waste utilize the platform of digital media to make late capitalist consumers aware of their responsibility towards the environment. This seemingly sudden cultural turn is not an isolated phenomenon but a direct consequence of the mainstreaming of the environmental problem in the contemporary moment. For example, the issues of pollution, global warming, and their immediate effects are no longer relegated to the distant future but are manifestly visible and drive policymakers to make informed decisions. In particular, media representation of New Delhi's continued poor air quality has been in the form of a "pollution emergency" (Safi 2017, n.p.). Similarly, drastic changes in weather patterns, an explosion of forest fires, and most recently resurgences of the Covid-19 pandemic have brought matters of ecological imbalance once again to the fore. Amidst these growing concerns, zero-waste lifestyle has emerged as an individual-centered method to address climate change by utilizing the virtual medium to reach out to disparate audiences and create global, sustainable communities. Zero-waste blogs/websites are, therefore, significant for this movement since they make the environmental message "fashionable" by using the online platform to popularize earth-conscious living.

Today, there are many advocates of zero-waste based in the Global South who take advantage of the wide reach of social media platforms to propagate sustainable living. For instance, Pankti Pandey is a zero-waste Instagram influencer who mobilizes environmental concerns by lobbying for "slow fashion" and following the principle of recycling and reduced consumption (D'silva 2023, n.p.). Slow fashion, as the term indicates, is a call for ethical production/consumption in the fashion industry in order to combat the burden of textile waste which is chiefly the result of overproduction. Similarly, Indian environmentalist Sahar Mansoor founded the e-commerce website "barenecessities.in" as a marketplace for daily use items, such as spa bars, handwash powders, moisturizers, and other household essentials that claim to be outside the plastic economy. The website minimizes the use of plastic and sells products packaged in glass bottles with wooden caps, except for handwash dispensers which use a plastic pump, and it provides shampoos in the form of a bar. The page further specifies guidelines on how to reuse the bottle after the product finishes and even provides the option to return and/or refill it. The website offers limited items in contrast to the plethora of choices that other e-commerce businesses sell, and this aligns with the minimalist principle underlined in its brand name, "Bare Necessities". Moreover, it also serves as an educational page spreading environmental awareness through board games, print, electronic media, talks, online courses, and workshops.

As the titles suggest, both "Trashisfortossers.com" and "Barenecessities.in" embody the principle of zero-waste culture and environmentally conscious living. Lauren Singer's webpage, by contrast, functions as an informative blog by addressing the primary barriers that might arise in initiating a zero-waste life and answering matters pertaining to style, beauty, and hygiene. She also provides DIY guides, reading lists, and documentaries to enlighten the visitor about the crisis of the Anthropocene. From blogs entitled "Guide To: Sustainable Laundry" to articles titled "Style and Sustainability: Can They Coexist?", Singer engages with the myriad reasons for which zero-waste culture is dismissed, whether this be the inconvenience it causes, lack of personalization, or the sheer unavailability of package-free goods (Singer 2023, n.p.). To resolve the problem of unavailability of either plastic-free products or items free from plastic or paper packaging, one of the tabs on the webpage titled "Shop Package Free" redirects the viewer to another website that serves as a one-stop marketplace for zero-waste living. Started by Singer, "Package Free" counters the ubiquity of plastic composition and packaging in modern living by providing alternatives to plastic bottles, bubble wrap and paper-packaging when supplying cleaning essentials, beauty, sanitary, or health supplies ("Package Free" 2023, n.p.). It sells compostable wooden brushes, laundry detergents without microplastics, steel straws, and storage containers. From DIY guides on composting biowaste to promoting the purchase of reusable, second-hand items, both websites demonstrate how capitalist markets have hampered individual choices by making households entirely dependent upon plastic cultures. The supermarket/hypermarket quintessentially indicates how much the dependence on single-use plastic and paper has become normalized. Thus, zero-waste culture interrogates mass production and consumerism and responds to the problem through the alternative capitalism of eco-friendly, sustainable products, and packaging.

Zero-waste culture has surfaced against the backdrop of capitalist consumerism, uncontrolled production, and the threat to the planet's sustainability. One of the biggest environmental risks today, apart from the burning of fossil fuels, is the overreliance on non-biodegradable plastic polymers that pose an immense challenge to the problem of waste segregation, product recycling, and effective disposal. In popular understanding, the mechanism of recycling and repurposing is believed to be an effective remedy to the environmental problem. However, as Judith Enck and Jan Dell (2022) explain, the chemical "composition" of plastic and the high cost involved make it difficult to recycle the product. Moreover, varied plastics contain "different chemical additives and colorants that cannot be recycled together" (n.p.), which complicates the optimistic narrative of humans recycling themselves out of the environmental problem. Zero-waste politics has emerged against such a backdrop of belief in the short-term solution, namely that a few ecological measures will be enough to drastically control the impact of human damage in the age of the Anthropocene. Zero-waste living underlines that the problem of waste is fundamental to environmental degradation especially since most of the discarded waste ends up occupying landfills. Therefore, proponents of zero-waste living who share their environmental journey on digital platforms provide remedies for how to become less wasteful in day-to-day living, and create digital, global communities of individuals who are willing to alter their lifestyle for the benefit of the planet.

While the phrase "zero-waste" sounds like an anomaly, since the production of waste is fundamental to human existence, it has gained a subcultural, digital presence by challenging the mainstream market culture of unbridled, endless consumerism. In simplistic terms, zero-waste living implies that the individual or household produces little to zero non-biodegradable waste and leads a plastic-free, ecofriendly life by refusing and reducing wasteful consumption. Such is the urgency of the climate change crisis and the need for sustainable measures that the United Nations on December 14, 2022 declared that 30 March should be observed each year as the "International Day of Zero Waste", defining zero-waste as "responsible production, consumption and disposal of products in a closed, circular system" ("International Day of Zero Waste" 2024, n.p.). Zero-waste websites also promote such a circular model by refusing plastic and paper-waste, since most of it ends up in landfills, and educating the public by creating a critical dialogue about environmental precarity that extends beyond nationalistic borders. Zero-waste environmentalism marks a radical departure from other ecological movements by adopting a bottom-up approach starting at the "back end of waste disposal" and then moving to the front ("Zero Waste Community Principles" 2024, n.p.). Hence, its aim of eradicating waste means it primarily targets non-biodegradable materials that cannot be conveniently put into a circular framework.

Zero-waste culture targets the normalization of plastic and paper for the sake of convenience. In doing so, its practitioners refuse to consume plastic products, or packaging and instead incorporate lifestyle changes. These may either be small measures such as carrying a cloth bag to avoid using plastic/paper bags, carrying reusable cutlery, travel mugs, and lunch boxes to avoid the use of single-use disposables, to more drastic measures related to health, beauty, and hygiene such as composting one's food waste, replacing tampons with silicone-based menstrual cups, thrift shopping, and leading a minimalist life to eliminate waste generation in the first place.

As effective as the zero-waste solution appears to be in addressing the crisis of environmental precarity, it is also a highly challenging mechanism for many. This is not least because of the difficulties it poses to everyday living, since plastic disposables are highly convenient, unbreakable, air-tight, and leak-proof – one reason for their enormous popularity – but mainly because it propagates a household-centric model of environmentalism. Controlling household waste and consumption practices alone cannot be efficient remedies when industrial waste forms a major component of the polluting forces. Moreover, given the unequal relations of power between nation-states and rising social inequality globally, a universal model for living a zero-waste life and reducing consumption seems unfair to the poorer sections of society (Chakrabarty 2016, 104). Social minorities such as Dalits have historically borne the burden of waste disposal and where they stand in the solution of "zero-waste" in India is also an important concern. Similarly, the "others" of society are already part of the "zero-waste" framework since their marginal position has denied them any luxury or extravagance. Dipesh Chakrabarty (2016) unequivocally makes the argument regarding global income and social inequalities by raising the pertinent question: "Should not the less developed and more populous countries (like China and India) have a greater right to pollute, while the developed nations take on more responsibility to make deep cuts in their emissions?" (104). It is indubitable from Chakrabarty's question that the history of modernization has not been universal and linear and, therefore, a prescriptive politics, burdening each individual with their environmental responsibility is deeply problematic, when a small number of nations has benefited more from a greater share of earth's resources than the rest of the world.

If zero-waste culture appears highly radical and unrealistic to implement in practice, then a crucial question to ponder is: Should the cultural politics of "zero-waste" websites be taken seriously when its environmentalism is consumption-driven, unfair, and often impractical? Zero-waste politics seems to deny the freedom that modernity bestows upon the individual by conceiving of consumerist decision-making in essentially moral terms. The refusal to buy selected items then acquires moral connotations indicating virtuous behavior and ethical living by contrast to the conduct of the less-privileged sections that cannot afford such principled living. Social and economic position shapes consumer decisions and complicates the subversive power that zero-waste politics assigns to consumer choices. Furthermore, zero-waste politics shifts the environmental burden from multinational corporations to end consumers and this simplifies the ecological crisis to become merely a matter of choice. Despite these many reasons which manifestly point to the apparent problems in the conceptualization and implementation of zero-waste politics, the concept still holds significance for interrogating capitalist modernity and its temporal progress as emblematized by the plastic economy, the consumption of mass-produced goods, and the ways they become an impediment to the survival of humankind.

Zero-waste ideology is significant in questioning the limits of the capitalist ethos that has provided humans with the "geological agency" responsible for planetary changes in the Anthropocene (Chakrabarty 2009, 208). According to Chakrabarty, it is industrialization that has initiated the planetary crisis (2009, 217). For Jason W. Moore (2016), however, the term "Capitalocene" is a more appropriate word choice in defining the relationship between capitalism and the natural environment than the term "Anthropocene": "Capitalocene signifies capitalism as a way of organizing nature – as a multispecies, situated, capitalist world-ecology" (6). Moore questions the "historical" trajectory of the Anthropocene in order to underline the centrality of capitalist forces in transforming the ecology (3). To follow on from Moore's argument, the historical factors that have contributed to the establishment of the Anthropocene as a dominant epoch require attention. Zero-waste movement is, therefore, important since it tackles the environmental problem by targeting late capitalist production and excessive consumption. Zygmunt Bauman (2004) famously argued that "[l]iquid modernity is a civilization of excess, redundancy, waste and waste disposal" (7). Seen in this context, the culture of zero-waste underlines that the problem of wastage is central to environmental precarity and cannot be resolved simply by sustainable measures alone. Instead, it requires restructuration of lives and eco-politics at a global scale.

One of the biggest deterrents towards realizing the goal of sustainable environment is the material plastic. Reducing the presence of plastics in food, soil, water, bodies, and even "human blood" (Carrington 2022, n.p.) is one of the greatest challenges humankind faces today. Rachel Carson's (1962) study *Silent Spring*, critiquing the harmful effects of pesticides and insecticides on biodiversity, human body, and soil composition, was path-breaking in drawing attention to how human interventions were threatening the balance of nature. In contemporary times, a similar engagement is required with the problematic resilience and durability of plastic – a material valued for the convenience it offers in diverse fields like medicine, technology, and consumer markets – and how such interdependence has resulted in threats to marine life, contamination of the soil, and the facilitation of what Rob Nixon (2011) terms "slow violence" (2). According to Nixon, "slow violence" is characterized by "delayed destruction that is dispersed across time and space, an attritional violence that is typically not viewed as violence at all" (2). Thus, the impending threat of consumable items rotting in landfills and the contamination of water bodies is a form of "slow violence" since the way they affect the ecosystem is gradual and not immediately visible (2). A telling example of the power of "slow violence" and the restructuring of ecology is the formation of "plastiglomerate" which, as Dana Phillips (2021) explains, is a type of rock formed by the combination of plastic and organic material after many years of decomposition (21).

The environmentalism of self-control

The zero-waste cultural shift has used the digital medium not only to connect with the wider community of environmentalists, but also to influence the public in matters related to sustainability through individual blogs, websites, and social media posts. Because of the immense popularity and mass appeal of the medium, zero-waste culture is certainly not exempt from the "attention-centric" economy of social media. The personal politics of environmentalism is inextricably linked to the culture of visibility and self-fashioning that social media perpetuates. Thus, the politics of zero-waste environmentalism can be interpreted by evoking Marshall McLuhan's (1994) famous phrase, the "medium is the message" (7), since the message of environmentalism appears to become overshadowed by the medium of digital technology. One of the negative consequences of such digital environmentalism is that the cultural visibility of the environmental activist becomes an end in itself in the free-floating, information-overloaded world of the internet. Ironically, the online visitor's uninterrupted consumption of the web of social media becomes one of the prerequisites for carrying across the message of zero-waste.

The popularity and visibility that social media offers complicate the environmental message of zero-waste culture and politics. In a world caught up in a social media frenzy of promoting veganism, fashion, fitness, skin care, travel, and lifestyle vlogs, zero-waste living seems to be just one of many fads. With its unique lifestyle of self-imposed restrictions in the choice of consumer products, their ingredients and packaging – to essentially reduce and eliminate paper and plastic waste – zero-waste culture promotes a life of minimalism, which almost acquires an "ascetic" dimension (Logan 2017, 600). Dana W. Logan (2017) has argued that whether it is Marie Kondo's popular "trend" of "decluttering" the wardrobe, or the minimalist lifestyle adopted by privileged elites, such popular practices are a form of asceticism which aims to reject "excess materiality" (600–601). However, their spiritual ascent to minimalist living is still not able to escape the material component since the decision-making involved with selective consumption about what product qualifies for the criteria and what is to be rejected "transfigure[s] the materiality of the body and the world, creating abundance out of absence" (2017, 608). Thus, objects gain a form of psychological significance that is far beyond their material value and utility. Zero-waste minimalism runs the risk of reifying some products in its primary choice of rejecting otherwise perfectly consumable items.

Furthermore, such is the cultural influence of social media that it has the invisible power to elevate any individual to the status of a celebrity. From food and travel vloggers to self-appointed experts in skincare, lifestyle, and finance, social media has manufactured public figures who use the platform for individual self-expression and visibility. The online video-sharing platform YouTube has given rise to self-proclaimed experts reviewing cinema, politics, and popular culture. While their influence is immediately democratizing since it blurs the divide between the intellectual/artist and the public, such celebrityhood, however, is rarely distinct from conventional cultural icons. Alice E. Marwick (2015) has contended how social media celebrities "emulate the visual iconography of mainstream celebrity culture" (139) to acquire attention. With its unconventional living, the zero-waste cultural trend makes for an interesting recipe to unleash the power of visibility, attention, and image-making on its middle-class audience. Hence, the vlogger documenting her zero-waste life consciously partakes in crafting a narrative that establishes her as an ethical, environmentally conscious subject, apart from rendering her social recognition. Such self-guided, self-monitored personal politics also becomes a neoliberal solution that privileges individual freedom and remains oblivious to the socio-economic conditions that enable and support such choices. The politics of zero-waste culture seem inherently paradoxical since they highlight the subversive power of the household in refusing capitalist consumption and its detrimental impact on the environment, while also promoting ecological capitalism, which is carefully fashioned through environment-centric politics and aesthetics.

Zero-waste environmentalism of the digital age can best be understood as a new wave of environmentalism that is a direct output of "late capitalist" society. Marxist critic Fredric Jameson (1991) distinctively critiqued postmodernism for becoming a "cultural dominant" (4) since art and culture no longer have an independent identity but are appropriated by capital. Following from Jameson's argument, zero-waste culture can be interpreted as the environmental logic of late capitalism, since environmental politics is unable to separate itself from capitalist forces, whether this be through the technology-mediated presence or the alternative capitalism it promotes by serving as a marketplace for sustainable goods.

The environmentalism of zero-waste culture seems unique in comparison to other Green movements which questioned environmental degradation brought about by industrial development. Ramchandra Guha (2000), in his book *Environmentalism: A Global History*, traces the diverse, overlapping histories of environmental movements globally and divides them into two waves, the first phase being a direct response to industrialization and the second being a more formalized "intellectual" and political movement (3–4). Guha finds precursors to modern environmentalism in the moral and scientific argument made for protecting nature and using its resources cautiously (2000, 6). Taking Guha's argument forward, a case can be made for a third wave of environmentalism in the form of zero-waste culture, which challenges the late capitalist, consumptive culture of fast-fashion, single-use products, and non-biodegradables items like plastics to venture into an ethical, environmentally informed capitalism. If the first and second waves of environmentalism argued for the preservation and protection of the environment and its precarious subjects (such as tribes, Indigenous people, and environmental refugees) by speaking in the language of rights and policies, the third wave is more individual-centric in its approach. It is temporally located in the neoliberal present while being globally scattered through the digital space. What connects such otherwise disconnected zero-waste practitioners from the Global North and the Global South is their ideology of environmentalism which believes that by altering practices of consumption the crisis of sustainability can be resolved. Thus, the overt activism of early environmentalism has made way for a form of digital activism that is less policy-driven and leans more towards the subversive power of the individual in the postmodern world.

Conclusion: poetics of zero-waste politics

Marxist criticism of the 20th century has exposed how late capitalist society is flooded with objects of consumption creating the illusion of an empowered subject when the real power lies with the object in question. From Fredric Jameson's (1991) critiquing of "postmodern culture" to Theodore Adorno's and Max Horkheimer's (1944) skepticism about "the culture industry", late capitalist society not only alienates the individual with objects of consumption but also commodifies art and culture. Amidst such appropriating tendencies of global capitalism, the rise of zero-waste websites and their optimistic belief in the subversive power of personal politics, although encouraging, is also highly paradoxical.

Zero-waste websites and social media accounts present a carefully manufactured aesthetic with the effect that the symbolic value of their products far exceeds their utilitarian purpose. The products available for sale signify a virtuous environmentalism, since they have been carefully curated into "Personal Care Kits", "Lifestyle and Travel Kits", and "Gift Boxes" for the conscientious consumer willing to buy their way into the environmentalist philosophy (Mansoor 2023, n.p.). These examples denote the limits to the eco-capitalism of zero-waste websites since their environmental politics is ultimately co-opted into the capitalist logic making them no different from those they critique. They also foreground the centrality of aesthetics as an essential component in zero-waste politics. Both websites represent environmentalism through a carefully chosen minimalist aesthetic that privileges the goodness of the natural and sustainable, and invites the customer to invest in the product as well as the philosophy.

Both "trashisfortossers.com" and "barenecessities.in" engage with the crisis of the Anthropocene and the drastic measures required for sustainability through beauteous visuals. The visual-aesthetic of sustainable cloth bags, glass jars, steel lunchboxes, biodegradable bamboo toothbrushes, and wooden cutlery juxtaposed with fresh vegetables and fruits form the minimalist aesthetic of zero-waste living. Ironically, the horrors of climate change, its apocalyptic threat visible in the form of loss of species, melting icecaps, and rising temperatures, are addressed through aesthetic sensibilities for the bourgeois audience. The representation of rotting waste, beached whales, the havoc of natural disasters, and the polluting nature of segregating waste challenges individual sentiments to evoke disgust, anger, and fear. Hence, the narrative of climate change is that of abjection (Kristeva 1982, 4), i.e., expelling that disturbs "identity, system and border", particularly since the impending threat is always present, lurking nearby, inevitable, and almost uncontrollable. The Covid-19 pandemic is a potent example of the abjection of the environment since the epidemic challenged the system of language and defied human reasoning to multiply in numbers through symptomatic/asymptomatic carriers and continued to oppose the faculties of meaning. Zero-waste politics does not engage with such a narrative of abjection precisely because that would entail acceptance of the negative view that environmental precarity challenges human rationality and cannot be conveniently resolved through lifestyle-based choices. It is, therefore, not surprising that the politics of zero-waste is conceptualized in aesthetic terms, imagined through a stylized neoliberal subject who embraces self-reflexive living and mindful choices.

Given the way modern capitalism and its consumptive ethos engender environmental crises, this chapter has focused on the counterculture of the zero-waste lifestyle, propagated by blogs and e-commerce websites in the Global North and the Global South, to assess its politics, aesthetics, and efficacy in battling the environmental problem. The digitally mediated environmentalism of zero-waste targets middle-class values, promotes ethical living through minimalist aesthetics and ideologizes individuals into believing that their self-righteous attitude can make a substantial difference. Individuals become active agents, capable of difficult choices rather than passively responding to the "calling" of capitalist objects. However, such a self-centric model of environmentalism is essentially trapped in neoliberal optimism and is located firmly within social media influencer culture and the validation it offers to the subject. Both the websites analyzed in this chapter, although belonging to two diverse cultural regions, appear identical in their environmental ideology and aesthetics. Such has been the homogenizing impact of global capitalism that zero-waste digital movement in the Global South and Global North becomes fundamentally similar in propagating a capitalist-centric model of environmentalism.

From an optimistic perspective, zero-waste politics makes households aware of the environmental problem and empowers them to take decisions that may ultimately push capitalist forces to change their modus operandi. However, such a hopeful conclusion appears more like wishful thinking than an attainable goal since it is not backed by policy-driven action but conceived as a solipsistic vision. In the year 2022, India banned the sale and distribution of single-use plastics in a landmark decision to curb plastic waste. However, even the imposition of a national ban has been ineffective in eradicating plastic pollution from land and water bodies. As Kushagra Dixit and Priyangi Agarwal (2023) observe, despite the ban, "single-use plastics [...] continue to flood markets in the city" (n.p.). Seen in this context, zero-waste cultural politics is a welcome step for revealing the interconnectedness between capitalist progress and the threat to ecological sustainability. While the self-centric model of zero-waste environmentalism is problematic, the politics advocating for radical measures are not dissimilar from international agreements which also underline the need for extreme measures to curb the problem. For instance, Article 2 of the Paris Agreement (2016) clearly articulates the need to control "the increase in the global average temperature to well below 2°C above pre-industrial levels" (n.p.). Zero-waste cultural politics may not be an entirely effective solution to the problem of environmental precarity, because being a product of the tech-savvy, late capitalist order, such environmentalism is caught up in its own poetics of self-fashioning. However, it is still a crucial subcultural politics that foregrounds how the problem of waste generation is essentially a capitalist problem.

References

Anon. n.d. "23 Zero Waste Blogs that Will Change Your Life." *Sustainable Jungle*. Accessed April 10, 2023. https://www.sustainablejungle.com/zero-waste/inspiring-zero-waste-blogs/

Adorno, Theodor W. and Max Horkheimer. (1944) 2002. "The Culture Industry: Enlightenment as Mass Deception." In *Dialectic of Enlightenment: Philosophical Fragments*, edited by Gunzelin Schmid Noerr. Translated by Edmund Jephcott, 94–136. Stanford, CA: Stanford University Press.

Baudrillard, Jean. 1998. *The Consumer Society*. London: Sage Publications.

Bauman, Zygmunt. 2004. "Culture of Waste." In *Wasted Lives: Modernity and its Outcasts*, by Zygmunt Bauman, 94–133. London: Polity Press.

Carrington, Damian. 2022. "Microplastics Found in Human Blood for First Time." *The Guardian*. March 24. https://www.theguardian.com/environment/2022/mar/24/microplastics-found-in-human-blood-for-first-time

Carson, Rachel. 1962. *Silent Spring*. Greenwich, CT: Crest Books.

Chakrabarty, Dipesh. 2009. "The Climate of History: Four Theses." *Critical Inquiry* 35 (2): 197–222.

Chakrabarty, Dipesh. 2016. "Whose Anthropocene? A Response." In *Whose Anthropocene? Revisiting Dipesh Chakrabarty's "Four Theses*," edited by Robert Emmett and Thomas Lekan. *RCC Perspectives: Transformations in Environment and Society* 2: 103–113. https://doi.org/10.5282/rcc/7421

Connett, Paul. 2013. *The Zero Waste Solution: Untrashing the Planet One Community at a Time*. White River Junction, VT: Chelsea Green Publishing.

Crutzen, Paul J. 2006. "The 'Anthropocene'." In *Earth System Science in the Anthropocene*, edited by Eckart Ehlers and Thomas Krafft, 13–18. Berlin: Springer.

Dixit, Kushagra, and Priyangi Agarwal. 2023. "Single-use Plastic Ban Stays on Paper." *Times of India*. July 24.

D'silva, Sharon. 2023. "Pankti Pandey: A Leading Voice in Zero Waste Living." *Femina*. October 17. https://www.femina.in/trending/achievers/pankti-pandey-a-leading-voice-in-zero-waste-living-276968.html

Enck, Judith and Jan Dell. 2022. "Plastic Recycling Doesn't Work and Will Never Work." *The Atlantic*. May 30. https://www.theatlantic.com/ideas/archive/2022/05/single-use-plastic-chemical-recycling-disposal/661141/

Gandhi, M. K. (1909) 1997. *'Hind Swaraj' and Other Writings*, edited by Anthony J. Parel. Cambridge: Cambridge University Press.

Guha, Ramchandra. 2000. *Environmentalism: A Global History*. London: Longman.

Jameson, Fredric. 1991. *Postmodernism or The Cultural Logic of Late Capitalism*. Durham, NC: Duke University Press.

Kristeva, Julia. 1982. *Powers of Horror: An Essay on Abjection.* New York: Columbia University Press.
Logan, Dana W. 2017. "The Lean Closet: Asceticism in Postindustrial Consumer Culture." *Journal of the American Academy of Religion* 85 (3): 600–628.
Mansoor, Sahar. n.d. "Bare Necessities." Accessed December 15, 2023. https://barenecessities.in/
Mauch, Christof. 2016. "Introduction: The Call for Zero-Waste." *RCC Perspectives* 3: 5–12. http://www.jstor.org/stable/26241370.
Marwick, Alice E. 2015. "Instafame: Luxury Selfies in the Attention Economy." *Public Culture* 27 (1): 137–160.
McLuhan, Marshall. 1994. *Understanding Media: The Extensions of Man.* Cambridge, MA: MIT Press.
Moore, Jason. W. 2016. "Introduction: Anthropocene or Capitalocene? Nature, History, and the Crisis of Capitalism." In *Anthropocene or Capitalocene? Nature, History, and the Crisis of Capitalism*, edited by Jason W. Moore, 1–11. Oakland, CA: PM Press.
Nayar, Pramod K. 2019. *Ecoprecarity: Vulnerable Lives in Literature and Culture.* New York: Routledge.
Nixon, Rob. 2011. *Slow Violence and the Environmentalism of the Poor.* Cambridge, MA: Harvard University Press.
"Package Free." n.d. Accessed December 15, 2023. https://packagefreeshop.com/
Palmer, Paul. n.d. "The Faux Zero Waste Movement Is Spreading." *Zero Waste Institute.* https://zerowasteinstitute.org/wp-content/uploads/2013/03/Faux-Zero-Waste.pdf
Phillips, Dana W. 2021. "'Paper or Plastic?' and Other Conundrums of Environmental Change." In *Life in Plastic: Artistic Responses to Petromodernity*, edited by Carren Irr, 11–30. Minneapolis, MN: University of Minnesota Press.
Safi, Michael. 2017. "Delhi Doctors Declare Pollution Emergency as Smog Chokes City." *The Guardian.* November 7.
Singer, Lauren. n.d. "Trash is for Tossers." Accessed December 15, 2023. https://www.trashisfortossers.com/
Standing, Guy. 2011. *The Precariat: The New Dangerous Class.* London and New York: Bloomsbury.
The Paris Agreement. 2016. *United Nations Framework Convention on Climate Change (UNFCCC).* https://unfccc.int/sites/default/files/resource/parisagreement_publication.pdf
United Nations. n.d. "International Day of Zero Waste." Accessed January 8, 2024. https://www.un.org/en/observances/zero-waste-day
Zero Waste International Alliance. n.d. "Zero Waste Community Principles: Global Principles for Zero Waste Communities." Accessed January 8, 2024. https://zwia.org/zero-waste-community-principles/

13 Hydrosophy

Ecology, choreography and multispecies precarity

May Joseph and Sofia Varino

Hydrosophy as praxis

May Joseph

Aquatopian performatives

Aquatopia, *The Hills, Governors Island, 2017. On a bright Sunday in the summer of 2017, a cast of over 20 performers converge in slow motion atop a stone hill at the southern tip of Governors Island. It is the damned year of five major hurricanes that have destroyed islands in the Caribbean and the Atlantic Seaboard. Gliding ritually across a radiating arc, Tibetan gongs, ektara and flutes in hand, the performers meet momentarily to the sound of Dante Alighieri's* Inferno *being recited in Italian by James Cascaito. A soprano, Renee Krylova sings Georg Handel's aria "Ombrai Mai Fu" and Sofia Varino recites Fernando Pessoa's poem "Mar Português". Swirling out of this momentary incantation sequence, over 50 people join the leaping, waving dancers climbing the spiral pathway up the mountain to the top, to meet an Odissi Dancer at the peak, purging the history of colonization from Lenapehoking. The winding weaving mass of humans holding blue, green and aquamarine fabrics fluttering in the wind swaddle the mountain, symbolizing the Hudson River's ever-present influence over the shape of New York's history. In* Aquatopia, *audience and performers blur, folding into one another as part of the landscape. The performance lasts an hour.*

In his book *Three Ecologies,* Félix Guattari ([1989] 2000) emboldens us to imagine and apply "a new ecosophy" (67). This new ecosophy is to be a multifaceted movement, a collective and singular subjectivity that exceeds individualization and stagnation. For Guattari, the new ecosophy is an immersive practice that should catalyze other kinds of worlds, new types of mental ecologies. Guattari's radical theory anticipates the unprecedented destabilization of human potentiality in the face of climate change (Figures 13.1 and 13.2).

DOI: 10.4324/9781003383598-18

Figure 13.1 Aquatopia – blue (2017). Photo courtesy of Joshua Kristal.

Figure 13.2 Aquatopia (2017). Photo courtesy of Joshua Kristal.

Drawing on this ontologically oriented rethinking of categories, praxis, and habits laid out by Guattari, the Harmattan theatre company founded by May Joseph in 2009 began a series of interactive immersions in the ecologies of hydrosophy, of water thinking, of water becoming, to engage with the challenges of rising seas and water levels. Spurred by Maude Barlow's (2007) terrifying question "where has all the water gone?" (9), Joseph assembled a performance lab of local dwellers based at different sites along the world's shorelines to explore the histories of water control and sovereignty (Barlow 2007, 37). Barlow's analysis of how the Global South has been forced into water privatization and how increasing water pollution in the Global South is convenient for the corporate interests in the Global North is one impetus to respond to the world's water crisis (36) through any means possible.

Along with core Harmattan performer and associate director Sofia Varino, Joseph created 23 site-specific durational performances with coastlines, buried water sources, and storm surge sites around the world. For Joseph, these engagements with the nonhuman are ecological assemblages, morphologies of movement and sentience that catalyze landscapes and bodies of water through movement. At the heart of these durational performances are the relationalities between the human, the nonhuman, and the atmospheric effects of wind, cloud, rain, heat, and water: a morphing, attenuating set of contestations, contemplations, and manifestations. Our performance praxis involves researching the interface of land and sea: historically damned or cursed coastal sites and environmentally damaged ocean sites that present their own narratives of environmental degradation. Water cycles, oceanic life, and what lies under the sea inform each Harmattan project engaging ontologies of the nonhuman. The sentient histories of sand, stone, ocean, seashells, rivers, drive our experiments in movement and shape our site-specific, large-scale live performances.

The questions shaping Harmattan's performances are speculative, framing the boundaries between the human and the nonhuman, between unlearning a human-centered worldview and relearning how to listen, move and see again, immersed in the nonhuman. Again, taking up Guattari's assertion that "[w]e need new social and aesthetic practices, new practices of the Self in relation to the other, to the foreign, the strange" and what Guattari identifies as "a constantly mutating socius", these new practices emerge from an "ethico-aesthetic discipline" produced by what Guattari calls "heterogenesis" ([1989] 2000, 47), the multi-sourced roots of speculative inquiry informing how a Harmattan project is conceived.

The collaborations between Joseph and Varino are practices in heterogenesis, a living praxis of hydrosophy. They are interrogations of the nonhuman through bodily, material, and sonic interactions with bodies of water, or hydrocorpora and water sites, or hydroscapes. Producing slow, durational movements at the interface of land and water, between stone and sound, on the thresholds of bridges and eroded bays, the improvisational, collaborative and ensemble-based performances are exercises in the multiple, transversal assemblages that Guattari identifies as "landscapes and fantasies of the most intimate spheres of the individual" ([1989] 2000, 47). These assemblages enmesh the human with the nonhuman, between an entrenched singularity and the radicalizing multiplicities of a multispecies becoming (Morton 2010). Performance opens the space of mutability between land and sea to allow for the more-than-human becomings of amphibious corporeality.

Re-enchanting climate change politics

Far Rockaway *performance, Tribute Park, Beach 116, Rockaways, September 12, 2013.*

A mournful sea chanty wafts across the salty sea air of Jamaica Bay. A group of performers walk slowly from a radial of directions towards Tribute Park. A harp joins in with Irish sea songs of desire and mourning. Elsewhere, a mellifluous saxophone rises in crescendo as an Odissi dancer clad in white dances ritually into the converging group of dancers who stand on large rocks strewn across the reconstructed promontory. The Odissi dancer enters the bay and submerges in its icy waters, looking out across the barrier islands. The performers also stand still, looking out towards the vast bay, the site of the last hold-out of the Lenape against the Dutch during the 1600s. This site-specific performance of one hour is done a year after Hurricane Sandy's devastation, at a site that was completely washed away by the storm. Performing at the restored park, which is a memorial to the 300 firefighters who died in the World Trade Center, and now a symbol of local resilience after the monstrous hurricane, is a complex enactment of Lenape land acknowledgement alongside the contemporary experiences of storm surge and environmental catastrophe.

Joseph's immersion in the hydrosophic potentiality of performance seeks to "re-enchant the world", as Silvia Federici (2019, 199) phrases it, through a praxis of "multifaceted movement" (Guattari [1989] 2000, 46) involving the perpetual surprise and destabilizations of a mutating socius of live music, dance, voice and choreographed slow walk. Taking a deindividuated minoritarian approach to imagining multispecies becoming, the choreography of Harmattan is drawn from what Federici refers to as "the loss produced by the long history of capitalist assault on our autonomous powers" (2019, 190). For Joseph, submergence in the ecologies of buried rivers and eroded bays is a way to open up what Federici refers to as "that complex of needs, desires, and capacities that millions of years of evolutionary development in close relation with nature have sedimented in us,

Figure 13.3 Far Rockaway (2014). Photo courtesy of Joshua Kristal.

which constitute one of the main sources of our resistance to exploitation" (190). For Federici, re-enchanting the world means to return to a sensory, kinetic, haptic engagement with its vitality: "I refer to our need for the sun, the wind, the sky, the need for touching, smelling, sleeping, making love and being in the open air, instead of being surrounded by closed walls" (190).

Federici's radicalizing call shapes the undercurrent of Joseph's directing and performance praxis. Performance is a laboratory to shock open the oppressions and forgettings of the bodies frozen in capital accumulation. Taking citizens of the city out into the open, down to their beaches in former apartheid landscapes, or across abandoned shorelines, animating neglected coastlines where the new questions of fragility and harm are intertwined with feelings of openness and rejuvenation, this is the new complex of needs and desires that Joseph's choreography brings to the fore. For Joseph, the praxis of hydrosophy offers a way to address what Federici calls "the great impoverishment" and to enable the "disaccumulation of our precapitalist knowledges and capacities" (2019, 191) (Figure 13.3).

Ocean morphologies

Dreamscapes performance, *Meg Webster's Stone Garden, Pier 57, Manhattan, September 11, 2010.*

A concert of large boulders from the Hudson Valley frames this performance about the Hudson River's shoreline. A group of 20 performers walk, dance, wander, meander, recite and meditate along the Hudson River's coast on September 11, 2010, nine years after the 9/11 catastrophe. They throw salt up in the air marking the old river line. The sun is intense and the dancers flow with harps, conchs, cymbals, a tambourine and percussion across the horizon of the garden to the end of the pier, embracing the river. The performance lasts an hour and the audience sits on hillocks watching the unfolding slow-motion ritual of movement and chant.

The oceans are already devouring large areas of the global land mass. We must draw upon the ocean for our practices of method, as Dilip Menon (2022) argues: "Thinking the ocean through the histories of migration begins to reckon with contiguous as much as continuous oceans rather than divided continents" (11). It is a historical necessity to comprehend fully the "multiple connections" that happen planetarily (11). Menon's point that "the ocean is a global entity" is critical to the work of Harmattan Theater's transmarine approaches (Figure 13.4).

Learning from the new rhythms, the transforming movements of the rising seas has necessitated Harmattan use a changing morphology of footwork, of human accountability with shorelines, teasing out the line "between Gaia and ground" to use Elizabeth Povinelli's (2021, 35) phrase. Harmattan's choreographic practice is a shattering of the "extimacies among water,

Figure 13.4 Dreamscapes (2010). Photo courtesy of Joshua Kristal.

person, place, ground" as Povinelli suggests, writing: "At the straining, extimate interface, new forms are coming" (35). Drawing on walking traditions from Kyogen, Kalaripayaatu, Chavittu Nadagam, Vipasana and Gandhian heritages, the emerging new forms of "extimacies", of evolving movements, take on distinctive characteristics depending on the specific landscapes and hydroscapes interfacing with performers: sea, sand, rocks, boulders, cliffs, bays, berms, winding pathways, spirals, hills.

Harmattan's experiments with the nonhuman are part of this search for new forms at the interface of humans, oceans and climate that also account for the histories of the disappeared. Its multi-sited and heterogenous ensemble is an enactment of the full capacity of the uses of democracy, where human actions are systems of reason, as Jacques Rancière (1995, 47) puts it. Performance is a living tool to examine Rancière's point that "the emancipated man is a man who walks and walks, moving around and conversing, putting meaning into circulation and promoting the movement of emancipation" (51). Again Rancière: "The democratic experience is thus one of a particular aesthetic of politics" (51). Extrapolating Rancière's arguments on the uses of democracy to understand the aesthetics of Harmattan Theater, the core principle is activating the interface between humans and the nonhuman in real time.

Planetary survival aesthetics

Harmattan theater was created as a speculative, amphibious tool to catalyze ecological thinking for a planetary survival aesthetics. What does a performative immersion with the more-than-human in relation to environmental precarity look like? Harmattan's part improvised, part virtuosic, open-ended choreographies are phenomenological experiments in what Judith Butler (2015) calls the performative theory of assembly (143). Butler points out that it is imperative that "bodies assemble", since "these forms of embodied and plural performativity are important components of any understanding of 'the people' even as they are necessarily partial" (8).

Drawing on Butler's theorizing of assembly, Joseph's notion of the ensemble is necessarily partial and incomplete, but an important materialization of active engagement with an environmental ethics at a time of neoliberal bluewashing. It is always unclear who "the people" are, but the will to gather, to not take the shoreline for granted and to publicly stage a submergence in aquatic and terraqueous ecologies, is a political act with all its uncertainties. This gathering is not virtuosic, as Butler highlights that physical assemblies imply mobilities (2015, 10). Neither is it necessarily "always a political good", as she points out (126). Butler observes that "[s]ometimes overcoming unwilled conditions of bodily exposure is precisely the aim of a political struggle. And sometimes deliberately exposing the body to possible harm is part of the very meaning of a political resistance" (126). It is this awareness that sometimes just deliberately exposing the body to the public eye becomes a political act of awakening to the praxis of hydrosophy, of water consciousness, of water politics, that is at the heart of Harmattan's praxis of hydrosophy.

Furthermore, this praxis of assembling along abandoned or precarious coasts is a process of reengaging with lost community histories through movement. Movement in the Harmattan praxis is a multi-mobility practice incorporating performers with varying degrees of mobility into the larger performance of submergence with the atmosphere, the terraqueous landscape. The dancer, the athlete, the frail, the faltering in gait, all converge in a Harmattan embrace of the shifting coastline through a performance of relational, interdependent living.

Every Harmattan performance merges the unplanned, random and open encounters between performers, spectators and nonhuman participants such as dogs, seagulls, sand crabs, bees and pigeons. Seriality, repetition and rhizomatic sequences become forms of assemblage shaping the new associations of assembly in a Harmattan performance. For Joseph, these prismatic morphologies of movement encapsulate what Donna Haraway (2008) evokes in her book *When Species Meet*. Haraway writes, "those who are to be in the world are constituted in intra- and interaction. The partners do not precede the meeting; species of all kinds, living and not, are consequent on a subject-and object-shaping dance of encounters" (4). Haraway's "ordinary beings-in-encounter" (5) become Joseph's Harmattan ensemble of humans and nonhumans, "entangled, co-shaping species of the earth", to use Haraway's term, a new praxis of heterogenesis, of inventing ways to become sentient and responsive to the new realities of an anthropogenic planet.

Harmattan Theater is interested in the volatile agency of water, its surges and explosions, its depletions and toxifications. It draws on Timothy Morton's (2010) observation that "[t]he ecological thought realizes that all beings are interconnected. This is the mesh. The ecological thought realizes that the boundaries between, and the identities of, beings are affected by this interconnection. This is the strange stranger" (94). Action, dance, movement, choreography and song are interwoven into the mesh of trees, stone, sand, sea, light and species traces. There is no outside of the terraquatic hybrid landscape merging land and sea in a Harmattan performance. The audience is part of the assembly, a component of the assemblage, drawn into the movement, partaking in the performance and in the anthropogenic reality of a sinking planet.

Tkuweyo, Lenape for waves

The year 2024 marks 400 years of the genocide of the Lenape peoples of Lenapehoking (New York City). My dream is to create a cascade of waves down a hill at the site of the landing of the Dutch on Governors Island, where the first populations of the Algonquin tribe were enslaved and disappeared. I make performances from my dreams. This is a practice I have taken from the great Senegalese filmmaker Djibril Diop Mambety (1999), who once said: "I make films about my dreams" (n.p.). Mambety's statement has been a creative spur for my own artistic practice to engage with climate questions.

The dream I have about *Tkuweyo* or *Waves* is a grappling with a decolonial performance methodology. The assemblage of seriality, repetition and duration will weave an excavating thread around the accursed space of Dutch colonization on Governors Island. Sofia Varino will appear as Fernando Pessoa, reciting from *Mensagem* (message) in Portuguese, as a reminder of the maritime histories that shaped the historical past of the early 17th century. The challenge is always how to speak from and engage with a colonized space in a settler colonial culture. This space of unease and discomfort at the junctures of the anthropocenic and the anthropogenic histories of our time makes this performance a daunting undertaking. The edges of Governors Island are all built up, but if you look hard and listen long, the drowned and buried histories of the past hang suspended in the air like a ghost. *Tkyweyo* will be one small gesture towards a planetary survival aesthetics. Incomplete, partial and imperfect, but nevertheless an assembly of performers and spectators contending with our island's ecological histories.

Elemental (im)permanence

Sofia Varino

An aquatic assembly

In *Chaosmosis: An Ethico-Aesthetic Paradigm*, Félix Guattari proposes a new model for participating in the networked chaos of planetary life (1995). Guattari enunciates a kind of ecological consciousness that exceeds and precedes political action or scientific knowledge, although it may also encompass both. For Guattari, chaosmosis articulates a mode of immersion in the world as it is, using aesthetic experience to structure and enable such immersions. Guattari describes this sensory and cognitive interaction with the world as a kind of ecological immersion into chaotic, disordered "heterogeneous Universes" (66) that precedes the western schism between self and other, subject and object, inside and outside, or indeed the hierarchical segregation of life forms into discrete species. For Guattari, this pre-linguistic state that anticipates the coming awareness of an individual self, separated from the world, is not a developmental phase but rather a state that is contiguous with and foundational to all future developmental phases into adulthood. Such sensory immersion in the living world with its myriad aesthetic possibilities is what Harmattan has aimed to cultivate in its performance experiments. By combining music and movement with text and landscape, our choreographies delineate a relational chaosmotic playfulness with our environ, a willingness to respond and adapt, to note and register, to intervene and interrupt.

In a Harmattan project, performance functions as a collaborative multispecies effort, affording a way to physically, cognitively and affectively interact with the lived environment and its human and nonhuman actants, in both a spatial and a temporal sense. All set in New York City, the performances that we discuss in this chapter are experiments in multispecies conviviality as an artistic research method for decentering the human. The tangled species that performatively participate in our projects constitute what I call here an "aquatic assembly", in line with Joseph's engagement with the praxis of the ensemble in conjunction with Butler's analysis of assembly as concept and practice. In *Dreamscapes* (2010), *Far Rockaway* (2013) and *Aquatopia* (2017), and in our future performance *Waves* (scheduled for 2024), the aquatic assembly that gathers by the New York City riverside is a temporary, ephemeral and precarious one.

First experienced live and later recalled, the performances move from events embedded in the networked collective memories of performers and spectators to documents chronicled in our writing or recorded as digital media, uploaded as video files to the Harmattan Theater website (https://harmattantheater.com/) and social media. Dispersed and fragmented, specific scenes from each performance encapsulate an aquatic logic in their choreography, in their materiality and in their visuality. The scenic and cinematic aspects of each Harmattan project rely on the specificity of each site's offerings, by deploying its histories and aesthetics as the basis for each project.

In *Far Rockaway* (2013), a dancer immerses their body in the Hudson, gazing in their stillness at Manhattan's iconic skyline. In *Aquatopia* (2017), a group of over 20 performers carries and waves a trio of bright cloths, enveloping a rocky hill in a crisscrossed configuration of vivid colors. And in *Dreamscapes* (2010), the water bearers return water to the Hudson by using metal containers, while several performers draw water lines on the dust, sand and gravel in Meg Webster's Stone Garden at Pier 57. In the planned performance *Waves*, decentering the human will also entail gesturing towards the past and the foundational dehumanization of Native Americans in US history. As key sources of environmental knowledge for mitigating climate change, Indigenous epistemologies provide scientific, philosophical, literary and artistic insight into the networked relationality of human and nonhuman life. The crucial role that Native American communities have played politically in campaigning for environmental justice are a contemporary extension of these commitments, by interweaving historical struggles for sovereignty with urgent ongoing demands for Indigenous and environmental rights. *Waves* will be a project shaped by how history haunts the present of New York City, while grappling with the unpredictability of a precarious future.

As an environmental performance collective working sporadically and voluntarily whenever financial circumstances allow it, Harmattan has devised a strategy for remaining active with minimal funding and virtually no institutional support. We rely on what we already have and is possible for us to access: public spaces, public domain literary sources, historical archives, performance traditions, organic and inorganic matter, the elements and our own living, breathing bodies – which, in the case of many of us, may be lacking in formal training and technique, but are more than willing to improvise and to learn, to experiment and to experience the familiar cityscape anew. The obstacles are many, always. Obtaining permits for performing in public space is certainly a major one. Another is remunerating professional performers fairly for their skills and experience and compensating unprofessional performers for their time and effort. And finally, performing in the open for ephemeral, often indifferent or suspicious, passing publics, while depending entirely on the weather conditions, can be frustrating and strenuous, demanding vast reaches of creativity and generosity for anyone involved in a Harmattan project.

In spite of all this, we have persisted over the years and continue to plan future performances alongside our teaching, writing, research and multiple professional commitments. There is something about the call of the elements, the need for environmental contact with the open, the urge to be in community with human and nonhuman bodies, that keeps us coming back. Ultimately, I believe it is this ethical paradigm of openness, contact, vulnerability, and exposure that continues to enable our multispecies aquatic assembly as an environmental praxis – "incomplete, partial and imperfect", as Joseph puts it above (see p. 236).

Elemental performativity

As a multispecies assembly and ensemble, Harmattan creates relational work that is conceived and executed in close partnership with the elemental performativity of each site's ecosystems and life forms. Considering the scale of open-air theater from Indigenous and ancient Greek performance traditions to contemporary environmental art, spatiality itself, both in a geographical and geometrical sense, plays a crucial role in our dramaturgy. Encountering a site and finding a way to relate to its spatial agency is also a way of contending with the elemental performativity of the elements whose presence structures each Harmattan show, whether they are conspicuously present or imperceptible: the sun and stars, clouds and sky, air and wind, soil and ground, leave their impressions on each performer and spectator, and cannot be controlled by any of us. As we have often observed, it is this act of being in prolonged, sustained companionship with water, sky, and land during a Harmattan show that generates a sense of creative communion, which in turn reverberates in the gestural and literary elements of the performance, rendering movement and text as seemingly more-than-human, ritualized components. The slow, minimalist choreography conceived by Joseph is learned, practiced, rehearsed, and eventually performed in collaboration with the nonhuman actants already present long before our arrival on site. As performers, we carry nonhuman matter and agentic forces in our own bodies, and its uncanny and abject dimensions participate in each of our performances of elemental impermanence.

In *Becoming Human: Matter and Meaning in an Antiblack World*, Zakiyyah Iman Jackson (2020) seeks to decenter the figure of the "human" in its presumed white, male and able-bodied subjectivity in order to examine black cultural production. This is, as Jackson makes clear throughout, also an ecologically attuned gesture of engaging with nonhuman species, subjectivities and materialities, whose assumed ontological plasticity has served to justify the twinned logic of genocidal and ecocidal violences. In *When Species Meet* (2008), Haraway shows how interwoven species is with race and gender (and, I would add, with ableist normativity). Anti-specism movements offer an astute critique of western modernity's drive to classify life forms according to species characteristics, noting the arbitrariness of imposing an implicit hierarchy on organisms, or of assuming unifying characteristics in spite of genetic variability. In *A Billion Black Anthropocenes or None*, Kathryn Yusoff (2018) offers an incisive demonstration of the racist and colonial legacy shaping the idea of the "Anthropocene", offering a critique closely aligned with Jackson's astute analysis. Yusoff observes that since geological time is so much longer than historical time, anthropogenic contributions to climate change cannot be reduced to a human scale of individual or even collective agency. Rather, climate change is the result of a complex process of colonialism and industrialization that is based on the extraction of value from human and nonhuman resources, predicated on systematic anti-Black violence.

The unexpected morphologies of New York City (NYC), with its geological (de)composition across a grid of asphalt, brick and cement, lend themselves to prolonged, daily contact with the riverside, albeit not without effort – it is necessary to cross an imposing highway, and although the riverside is theoretically wheelchair accessible, the realities of navigating the NYC public transportation system make access to the riverside demanding for anyone with a disability. Accessing Governors Island from New York

City is not straightforward either, entailing a ferry ride in addition to a fair amount of walking, both time- and energy-consuming. The water bearers in *Dreamscapes* (2010) play with water and with salt to consider the convoluted flow of the Hudson in its perpetual dance with the ocean, retracing the original lines of the river by throwing salt into its waters. Our affective investments in the water are not just about survival of the species (our human species) but are rather shaped by experiential, sensory, embodied memories and attachments to the living world, the very world that enables (our human) life. These affective investments are mediated by acts of multispecies imagination, of reasoning, and of symbiotic exchange.

In Bruno Latour's (2005) influential Actor-Network Theory (ANT), an actant is defined as any entity with the capacity to act, influencing a process or course of events with its participatory agency. Centering the nonhuman actants we work with (from the Hudson to the Atlantic, to the sand, stones, and sky), elemental performativity can be understood as not only the lively capacity of matter to act, but also to remain inert, indifferent to our efforts as performers, relentless in its own volition. Performing in the open brings our human scale to the fore, highlighting the limits of human agency. Elemental performativity is thus my way of gesturing towards the unpredictable volatility of the living world's atomic, molecular structures as they unfold alongside us: sea, river, rock, gravel and grass, all that we touch as we move across the water-bound livespaces of a Harmattan show.

In spite of our efforts at decentering the human, the ensemble reminds us of how entirely human we really are in every sense: performers must be fed and keep themselves hydrated, and the spontaneous gathering around the vegan catering provided for a performance day is a central part of the sociality of each Harmattan show: we gather, we chat, we exchange ideas and updates, we make plans, we rehearse and improvise. We are bound to our limitations and discomforts, our pleasures and habits, our skills and experiences. We bring to the choreography our weary, clumsy bodies, our lack of formal training, our voices hoarse from trying to shut down the noises of city traffic, passers-by, the roar of the ocean, the trickle of the river. We are a gregarious ensemble, both hosts and guests, open to spontaneous conversations, encounters, and unexpected encounters with the inhabitants and passersby.

Affective ecologies

It is not by chance that ecosophy, which centers affective, embodied, and aesthetic experience, is informed by the radical psychiatry and anti-psychiatry movements. Guattari's ecosophy is a radical proposal for feeling, thinking, and being. Set on dismantling individual subjectivity in its capitalist, colonial and heteropatriarchal foundations, ecosophy eschews individualized (human) psychology to inaugurate an affective and aesthetic regime where relationality can bring subjectivity into an errant, volatile realm.

Sound marks a Harmattan performance in myriad ways. The roar of traffic and river mingles with the delicate flutters of birdwings and birdsong, the insistent, piercing cries of seagulls and the repetitive burst of horns. In the midst of this cacophony, the musical instruments played by several members of the company provide a unifying interruption, marking the arrival of another timespace. The Harmattan soundscape is also a multispecies, collaborative one. The literary sources in a Harmattan show come from western (pre)modernity, with its colonial logic justifying the violence needed to extract value from dehumanized bodies, impose reason and order on all life forms. From Dante's *Inferno* to Pessoa's *Mensagem*, these texts embody modern myths of ascension, progress and expansion, which serve to justify the atrocities of colonialism. The textuality of these works is delivered by the vocal cords of the human throat, a limited instrument in terms of its vocal capacity. Surrounding sounds drown out our words, and we do not use microphones or any kind of electrically powered tools. Out in the open, the precarity of human language, exposed in its oral, physical dimensions to the vicissitudes of sound and air conditions, produces a multispecies soundscape enacted in real time.

Jane Bennett (2010) refers to enchantment as a kind of "impersonal affect", defining it as "that strange combination of delight and disturbance" (xi). In a Harmattan project, the act of re-enchanting affective and sensory impressions is also one of reverence and resistance, experimenting with new ways of being and doing, considering how ephemeral and lush our daily New York City lives really are. When Harmattan performers begin to arrive at a performance cite in NYC, this gradual agglomeration of friends and strangers brings a heightened sensory awareness to the familiar spaces we inhabit. The provisional audiences that watch our shows, whether tourists or locals, can grapple with the elemental performativity of the hills in Governors Island, with the stones and hillocks in Pier 57, or with the Hudson's still, steady waters. We realize how agential, how scenic, how artful the sky can be when it illuminates the scene, and brings its own light scheme and heat pattern, determining whether sweat or rain will be challenging us. Walking slowly or holding still, we experience the direction of the wind and the waft of smells it carries, whether they are pleasing flowering ones, the metallic musk of industrial waste, or the putrid stench of decaying trash in the city streets on a hot summer day.

Arts of losing

"The art of losing isn't hard to master", observes Elizabeth Bishop ([1976] 1983) in her famous poem "One Art". The habitual exposure to news of environmental disaster has become part of our daily lives, as has environmental pathogens. The losses caused by climate change include material and economic as well as affective and social losses. The realities of rapid species extinction, habitat loss, deforestation, rising sea levels affect our lives to various extents according to our degree of socioeconomic privilege and geographical location. Although this violence is slow, gradual and perhaps imperceptible, it is also relentless, continuous and, to a large extent, unavoidable. Rob Nixon (2011) points out that since the effects of environmental violence are incremental but relentless, they can be difficult to detect and document. In line with the gradual processes of environmental violence, the slow walk of a Harmattan choreography is an enactment in real time of these often-imperceptible violences. Slowing down alters our perception and develops our "arts of noticing", as Anna Tsing (2015, 37) puts it. Precarious and vulnerable, our living bodies acquire an organicity in tune with the gradual, incremental duration of ecological processes around us.

In *Vibrant Death*, Nina Lykke (2021) considers mourning as a practice of alterity, of conversing and communing with the dead, of viscerally experiencing past memories and bonds, and of attunement to the vibrancy of death. I contend that it may be a crucial practice for grappling with the challenges of ecological devastation, from the human lives lost to pollution and climate change to the nonhuman life forms on the brink of extinction. We must cultivate arts of noticing and mourning, which are more aligned with passivity, inaction and reflection, alongside our arts of survival and endurance. This includes the capacity to acknowledge, with awareness and compassion, the plight of so many human and nonhuman lives erased due to relentless colonialism and industrialization.

Mercury, lead, plastic, debris and toxic waste contaminate the water and introduce unexpected variables into the delicate equilibrium of water-bound ecosystems, from the deep sea to the New York City shore or the Hudson riverside. But it is with this contaminated, multispecies body of water that we must collaborate, engaging with the biochemical (de)composition of its hydromorphologies. Carbon fuels pollute the air we breathe but have especially deadly consequences for those exposed to more saturated air. The connections between the health of human and nonhuman populations are foundational in Indigenous and traditional medicine, and first reached a wide audience in the US when Rachel Carson (1962) formulated this connection in western scientific language in *Silent Spring*.

Confronted with anthropogenic climate change, NYC communities are also faced with the limitations of political and technological intervention. Acts of mourning that engage with the city's histories, acknowledging its violent colonial past and precarious future, may enable another, less human-centered kind of activism. This may give rise to protests that are more contemplative than demanding, to campaigns that are less about policy change and more about critical reflection, less invested in solutions and more concerned about acknowledging uncertainty and doubt. Learning to converse with death and extinction may become an important way to remain responsive and sentient, connected to a planetary precarity that seems more and more to have become a precondition for the continuation of life.

In our performances, local species come alive in a way we can respond to and collaborate with as artists and as earthlings, inhabitants of a planet on the brink of mass extinction. This responsiveness inaugurates, I argue, an embodied ecological ethics that is as invested in political in/action as it is in experimenting with communal ties and social relations. Considering elemental performativity enables a way of being in the world that is more lively, playful, responsive, caring and capacious. That, at least, is the hope I share with May Joseph and with the Harmattan Theater ensemble.

References

Barlow, Maude. 2007. *The Blue Covenant: The Global Water Crisis and the Coming Battle for the Right to Water.* New York: The New Press.

Bennett, Jane. 2010. *Vibrant Matter: A Political Ecology of Things.* Durham, NC: Duke University Press.

Bishop, Elizabeth. [1976] 1983. "One Art." In *The Complete Poems 1927–1979.* New York: Farrar, Straus & Giroux.

Butler, Judith. 2015. *Notes toward a Performative Theory of Assembly.* Cambridge, MA and London: Harvard University Press.

Carson, Rachel. 1962. *Silent Spring.* Boston and New York: Mariner Books.

Federici, Silvia. 2019. *Re-Enchanting the World: Feminism and the Politics of the Commons.* Oakland, CA: PM Press.

Guattari, Félix. 1995. *Chaosmosis: An Ethico-Aesthetic Paradigm.* Translated by Paul Bains and Julian Pefanis. Bloomington and Indianapolis: Indiana University Press.

Guattari, Félix. [1989] 2000. *The Three Ecologies.* Translated by Ian Pindar and Paul Sutton. London and New Brunswick: The Athlone Press.

Haraway, Donna. 2008. *When Species Meet.* Minneapolis, MN and London: University of Minnesota Press.

Jackson, Zakiyyah Iman. 2020. *Becoming Human: Matter and Meaning in an Antiblack World.* New York: New York University Press.

Latour, Bruno. 2005. *Reassembling the Social: An Introduction to Actor-Network Theory.* Oxford: Oxford University Press.

Lykke, Nina. 2021. *Vibrant Death: A Posthuman Phenomenology of Mourning.* London and New York: Bloomsbury.

Mambety, Djubril Diop. 1999. "The Hyena's Last Laugh: A Conversation with Djibril Diop Mambety" (with N. Frank Ukadike). *Transition* 78. Reprinted in abridged form in *Metrograph.* Accessed May 24, 2024. https://metrograph.com/a-conversation-with-djibril-diop-mambety/

Menon, Dilip M. 2022. "Oceanic Histories: From the Terrestrial to the Maritime." In *Ocean as Method: Thinking with the Maritime*, edited by Dilip M. Menon, Nishat Zaidi, Simi Malhotra, and Saarah Jappie, 1–24. London and New York: Routledge.
Morton, Timothy. 2010. *The Ecological Thought*. Cambridge, MA: Harvard University Press.
Nixon, Rob. 2011. *Slow Violence and the Environmentalism of the Poor.* Cambridge, MA and London: Harvard University Press.
Povinelli, Elizabeth A. 2021. *Between Gaia and Ground.* Durham, NC: Duke University Press.
Rancière, Jacques. 1995. *On the Shores of Politics*. Translated by Liz Heron. London and New York: Verso.
Tsing, Anna. 2015. *The Mushroom at the End of the World.* Princeton, NJ and Oxford: Princeton University Press.
Yusoff, Kathryn. 2018. *A Billion Black Anthropocenes or None*. Minneapolis: University of Minnesota Press.

Index

Note: *Italic* page numbers refer to figures and page numbers followed by "n" denote endnotes.

For Product Safety Concerns and Information please contact our EU
representative GPSR@taylorandfrancis.com
Taylor & Francis Verlag GmbH, Kaufingerstraße 24, 80331 München, Germany

www.ingramcontent.com/pod-product-compliance
Ingram Content Group UK Ltd.
Pitfield, Milton Keynes, MK11 3LW, UK
UKHW022040061126
477531UK00009B/456
* 9 7 8 1 0 3 2 4 6 8 8 0 8 *